SKYGUIDE
A FIELD GUIDE TO THE HEAVENS

by
Mark R. Chartrand
Senior Vice President
National Space Society

illustrated by
Helmut K. Wimmer

GOLDEN PRESS • NEW YORK
Western Publishing Company, Inc.
Racine, Wisconsin

PREFACE

The sky is a part of nature available to all people, yet how the sky appears is very localized and very personal: No one sees the sky in exactly the same way or the same time as you do. This guidebook provides an introduction to the natural sky for those with little knowledge of astronomy, so that they can begin to appreciate the beauty of the sky and of the laws of nature that determine the motions and appearances of celestial objects. It is hoped that readers will find this book useful even when they have graduated to more technical, more detailed works.

The author wishes to thank many people who have played major and minor roles in bringing this book to the public. In particular, I am grateful to persons who have influenced my astronomical education at critical times: Arthur P. Smith of the Southern Cross Astronomical Society, Dan Snow of the Ralph Mueller Planetarium, Dr. Peter Pesch of Case Western Reserve University, and the late Dr. Sidney W. McCuskey, also of CWRU. Appreciation is due also to Barbara Williams, who initiated this project; to Jerome Wyckoff, who edited the text; and to Caroline Greenberg, who saw the book through to completion. The help, advice, and suggestions of my colleagues at the Hayden Planetarium in New York—particularly Allen Seltzer—have been most helpful. I am indebted to the facilities of the Perkin Memorial Library at the Hayden Planetarium. In a very special way, I must thank the artist for this book, Helmut Wimmer—co-worker, colleague, and friend.

The artist would like to thank, particularly, Barbara Williams for her confidence in his ability to visualize the book; also, Dr. Mark R. Chartrand for his assistance and guidance, and Ms. Sandra Kitt for her help in his research.

Mark R. Chartrand
Helmut K. Wimmer

Revised Edition, 1990

GOLDEN®, A GOLDEN FIELD GUIDE, and GOLDEN PRESS® are trademarks of Western Publishing Company, Inc.

TABLE OF CONTENTS

INTRODUCTION

> Look up at the bright and unsullied hue of heaven and the stars which it holds within it, wandering all about, and the moon and the sun's light of dazzling brilliancy: if all these things were now for the first time, if, I say, they were now suddenly presented to mortals beyond all expectation, what could have been named that would be more marvelous than these things, or that nations beforehand would less venture to believe could be? Nothing, me thinks: so wonderous strange had been this sight. Yet how little, you know, wearied as all are to satiety with seeing, any one now cares to look up into heaven's glittering quarters!

This quotation could very well describe the present common disregard for the sky, yet it was written around 55 B.C. by the scholar Lucretius in his great work *De Rerum Natura.* Some things just don't change.

The sky is that part of the natural environment which we all share, no matter where we are on Earth—or even in space. Yet many people who can identify birds or rocks, trees or ferns, cannot point to any constellations by name, or find the planets in their wanderings among the stars. Few can tell why the Sun rises where it does or why the Moon has phases. This book is for them, and for all others who have ever wondered at the beauty of the night sky, watched the solemn progress of a lunar eclipse, or stood breathless for the timeless moments of the totality of a solar eclipse.

Lucretius refers to "the bright unsullied hue of heaven," and here he had an advantage over us. Modern people live, for the most part, under canopies of polluted air and urban lighting; these cut us off from the crystal clarity of the night sky seen in areas far from cities. But there is still much to be seen, even from cities.

If you can, get to a park, beach, or other area as far as possible from cities. Campers in the Southwest probably enjoy the best skies, but merely getting 50 miles from a city can be a great help. But be warned: If you have learned your constellations in a city (which is possible, at least for the brighter ones), you may get lost in the skies of the country. There are so many more stars! They were there all the time, of course, but in the city they couldn't be seen.

Another important factor in viewing, especially with a telescope, is what astronomers call "seeing." This refers to the steadiness of the images, rather than the transparency of the sky. If the stars are twinkling a lot, this is poor seeing, and a telescope will only exaggerate the poor images further. All objects in the sky appear to twinkle when near the horizon because their light is passing through more atmosphere before it reaches your eyes. With good seeing, objects high in the sky will seem steady. Ironically, some nights when there is just a bit of haze, and the transparency is not perfect, are the times of best seeing, at least in cities, because air turbulence is not distorting the image. Astronomers reduce this problem by placing their major instruments high atop mountains in carefully selected locations to get above as much of the atmosphere as possible.

Throughout this book references are made to "amateur" instruments. This is not meant to disparage, but simply to refer to telescopes and other equipment that are within the range of the nonprofessional stargazer. It definitely does not imply inferior quality, for many small telescopes are of high quality.

Astronomy is one of those rare sciences in which the serious amateur can make important contributions. Here, "amateur" means only "not getting paid for it." Yearly, amateur astronomers contribute over 100,000 observations of variable stars, discover new comets, watch grazing lunar occultations, discover novae and supernovae, and count sunspots. All these add to the store of astronomical data from which professional astronomers draw to check their theories and calculations. In few fields can the amateur make such contributions.

You should try to make contact with a local amateur astronomical society. Many hold regular meetings with speakers, and occasionally have "star parties" at which many telescopes are set up to observe the sky. Here you will have an excellent opportunity to compare many instruments, observe many astronomical sights, and make friends.

Local science museums and even some public libraries often have telescopes available for public viewing on a regular basis, as well as introductory classes on astronomy. To learn your way around the night sky, there is no better place to start than your local planetarium, where the sky and the weather can be controlled. If you are traveling anywhere near one of the major observatories, you should try to plan to stop in on one of their public nights. A list of many of the major observatories and planetariums can be found on pp. 265-266.

A word of advice about buying telescopes: Decide first if your interest in the subject is serious and permanent, or if you just wish an occasional view of the sky. For a first instrument, binoculars are ideal. They are less expensive than a telescope of comparable size, easy to hold, have a wide field of view, and, if your interest wanes, can always be used for watching sports or birds! After you have had a little experience, ask local amateur astronomers for recommendations on what to buy, after reading pp. 30-39. Some local amateur astronomical societies even offer used telescopes for sale.

A second word of advice about buying telescopes: Avoid any telescopes sold in department stores or by opticians. The telescopes are almost invariably of poor quality and not worth the price, and the salespeople know little or nothing about them. There are only a few shops around the country specializing in amateur astronomy equipment, but these are easy to identify. By and large, the better amateur instruments are available from mail-order firms (see p. 269).

When you first look through a telescope at, say, Mars or Saturn, or the Ring Nebula, don't expect it to look like the photographs you see in many astronomy books. These photos were taken through the world's largest and best instruments. Your 6-inch reflector is not going to yield an image of comparable size and quality. Your telescope excels, however, in giving you a view that is particularly yours, is real, and is now. That intangible quality is something you can't get from staring at a picture.

Lastly, remember that astronomy is unique in that you can appreciate it on many levels, from just enjoying the sparkling beauty of the night sky, to keeping track of variable stars, to learning relativistic astrophysics. The depth to which you go is up to you. And although the same stars will appear in your skies in the same places year after year, all your life the planets will keep shifting in the heavens, comets will come and go, meteors will flash ephemerally across the sky, and new discoveries will be made, adding depth to your understanding of the old familiar sky.

Good observing!

ASTRONOMICAL SYMBOLS AND ABBREVIATIONS

SUN, MOON AND PLANETS

☉ The Sun	♀ Venus	♃ Jupiter	♆ Neptune
☾ The Moon	⊕ Earth	♄ Saturn	♇ Pluto
☿ Mercury	♂ Mars	♅ Uranus	

SIGNS OF THE ZODIAC

♈ Aries	♋ Cancer	♎ Libra	♑ Capricornus
♉ Taurus	♌ Leo	♏ Scorpius	♒ Aquarius
♊ Gemini	♍ Virgo	♐ Sagittarius	♓ Pisces

TERMS INDICATING RELATIVE POSITION

☌ Conjunction	☐ Quadrature	☊ Ascending Node
☍ Opposition		☋ Descending Node

ABBREVIATIONS

h, m, s—Hours, minutes, seconds of time
° ' "—Degrees, minutes, seconds of arc

A.U. or a.u.—Astronomical units
α or R.A.—Right ascension
δ or Dec.—Declination
lt-yr—Light-year

THE GREEK ALPHABET

A, α	Alpha	H, η	Eta	N, ν	Nu	T, τ	Tau
B, β	Beta	Θ, θ	Theta	Ξ, ξ	Xi	γ, υ	Upsilon
Γ, γ	Gamma	I, ι	Iota	O, o	Omicron	Φ, φ	Phi
Δ, δ	Delta	K, κ	Kappa	Π, π	Pi	X, χ	Chi
E, ε	Epsilon	Λ, λ	Lambda	P, ρ	Rho	Ψ, ψ	Psi
Z, ζ	Zeta	M, μ	Mu	Σ, σ	Sigma	Ω, ω	Omega

TERRESTRIAL COORDINATES

How you see the sky depends upon where you are on Earth. Even the way the sky seems to move overhead depends upon your terrestrial location. To specify a location on Earth precisely, two coordinates are used—two because Earth's surface is two-dimensional.

Latitude is the north-south coordinate, measured in degrees north and south of the equator, which is the imaginary line around our planet midway between the poles. The poles are the ends of the planet's axis of rotation, and for our purposes can be considered fixed. The equator is at latitude 0°. Northern latitudes are positive, southern ones negative. The north geographic pole is at latitude +90°, the south geographic pole at −90°. Locations having the same latitude are said to be on the same *parallel of latitude*.

Longitude is the east-west coordinate. A line from one pole to the other is called a *meridian of longitude*. (Such a line is an example of a *great circle*: a line on the surface of a sphere which is part of a circle whose center coincides with the center of the sphere.) A *prime meridian* is needed from which to start measurements eastward or westward. A century ago, different prime meridians were in use by different countries, each wanting the prime meridian to go through its capital. Since that situation was confusing to navigators and astronomers, all nations finally adopted the present prime meridian, which runs through Greenwich, England. Sometimes called the *Greenwich meridian*, it is marked at the Greenwich Observatory by a brass line in the pavement, so that you can stand with your feet in different hemispheres.

Longitude, like latitude, can be measured in degrees—from the prime meridian east and west up to 180°. At 180°, opposite the Greenwich meridian, is the *international date line*, which divides the Pacific Ocean. (For political and practical reasons, the line zigs and zags near the 180° meridian.) Along this line each new day begins and the previous day ends, so that, for example, when the time is 12:01 a.m. Wednesday just east of the line it is 12:01 a.m. Thursday just west of the line. Since the time of day when we see astronomical objects depends on Earth's rotation, longitude is often measured in hours, one hour equaling 15°. (See pp. 16-17.)

Although Earth's surface is two-dimensional, the planet is three-dimensional. Latitude and longitude are not just lines on the surface, but refer to angles measured at Earth's center. Latitude is the angle between the equator and the location referred to, and longitude is the angle between the prime meridian and the meridian of the location referred to. Each degree of latitude or longitude can be divided into 60 minutes (abbreviated ′), and each minute into 60 seconds (″). But navigators and astronomers commonly use degrees with decimal fractions, rather than minutes and seconds, because they make computations easier.

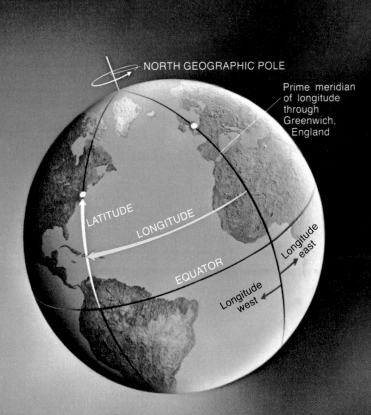

NORTH GEOGRAPHIC POLE

Prime meridian
of longitude
through
Greenwich,
England

LATITUDE

LONGITUDE

Longitude
east

EQUATOR

Longitude
west

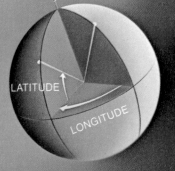

NORTH GEOGRAPHIC POLE

LATITUDE

LONGITUDE

HORIZON COORDINATES

Looking up at the sky from a plain, we seem to be at the center of the universe, with a flat Earth stretching off to infinity in all directions. We are at the center of the *celestial sphere*, an imaginary globe of indefinitely large size, on the inner surface of which all heavenly bodies and their motions appear as if projected onto a screen. This concept allows us to ignore *distances* of celestial objects except when relevant, heeding only their *directions*.

Often an observer wants to specify the direction of a celestial object with respect to *his* location. One convenient way is by means of *horizon coordinates*, sometimes called *topocentric coordinates*.

First, imagine a flat plane tangent to Earth's surface at your location. This is your *horizon plane*. The line along which it intersects the celestial sphere is your *horizon*. This divides the universe into the part you see above the horizon, and the invisible part below it. North is the point on the horizon in the direction of the geographic north pole; south is opposite. Facing south, east is on your left, west on your right.

Your celestial sphere has a spot unique for you: the point directly overhead, the *zenith*. If you move, your zenith moves. No other location on Earth has the same zenith as yours. The point opposite the zenith on the celestial sphere is the *nadir*.

The location of a celestial object with respect to the observer's location is described by the two angles called altitude and azimuth.

Altitude is the angle between the horizon and the object. This angle is measured from the horizon and perpendicular to it. Altitude is not height above the ground, as of an aircraft. It is an angle centered on you; or, an arc along the celestial sphere. Altitude ranges from 0° at the horizon to +90° at the zenith. Objects below the horizon have negative altitudes. Objects at the same altitude are said to be on a *parallel of altitude*. A line of constant altitude around the sky is called an *almucantar*.

Azimuth is the compass direction toward an object. More precisely, it is the angle measured around the horizon, beginning at north (0°) through east (90°) and so on until you reach a line drawn perpendicular from the object to the horizon. Thus east is azimuth 90°, south 180°, west 270°, and north 360° — the same as 0°.

You can estimate altitude or azimuth simply by extending your fist to arm's length. At this distance your fist will appear to be about 10° wide. If you align the bottom of your fist with the horizon, the top is at about 10° altitude. Stacking fist on fist, you can estimate wider angles. This method of approximately measuring distances across the sky is convenient when you are attempting to locate a star or other object that is a known number of degrees from an object you have already identified. A navigator uses a sextant to determine altitude and azimuth accurately.

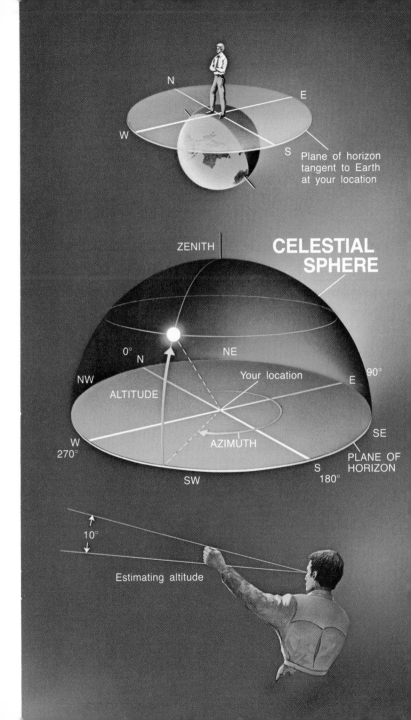

Plane of horizon
tangent to Earth
at your location

ZENITH

**CELESTIAL
SPHERE**

0°
N

NE

NW

ALTITUDE

Your location

E 90°

W
270°

AZIMUTH

SE

PLANE OF
HORIZON

SW

S
180°

10°

Estimating altitude

THE CELESTIAL SPHERE

Imagine Earth as surrounded by a very large sphere on the surface of which all stars, planets, and other celestial objects are placed. This is similar to the observer's celestial sphere mentioned on p. 10, but not exactly the same. Now we are imagining the astronomer's celestial sphere, with the center of Earth as the center of the sphere.

Next, imagine Earth's equator expanding outward until it intersects the sphere. The intersection will be a great circle on the sphere called the *celestial equator*. It is halfway between the *north celestial pole* (NCP) and the *south celestial pole* (SCP), which are the locations on the celestial sphere directly above the north and south geographic poles on Earth's surface.

In the sky there are counterparts to meridians of longitude, but for reasons explained more fully on pp. 16-17, they are usually called *hour circles*. You should keep in mind that as Earth turns, the meridian of longitude under a given hour circle changes continuously.

Another fundamental plane in the sky is the plane of Earth's orbit, called the *ecliptic plane*, or *ecliptic*. As seen by us on Earth, the ecliptic is an imaginary line in the sky along which the Sun moves during a year's journey around the sky against the background of stars. As seen by an observer sitting "on the NCP," the Sun would move counterclockwise. Perpendicular to the plane of the ecliptic are the north and south *ecliptic poles*.

When Earth formed from a cloud of gas and dust about five thousand million years ago, its axis of rotation did not coincide with the axis of the ecliptic poles. The two axes, and hence the plane of the equator and the plane of the ecliptic, are inclined to one another at an angle of about 23½°. This angle is called the *obliquity of the ecliptic*. Thus the north ecliptic pole is at R.A. 18h, Dec. +66½°, while the south ecliptic pole is located opposite at R.A. 6h, Dec. −66½°.

The two points on the celestial sphere where the equatorial and ecliptic planes meet are termed the *equinoxes*. The point at which the Sun crosses the celestial equator when going north in spring is called the *vernal equinox*. The point where the Sun crosses when going south in autumn is the *autumnal equinox*. (See pp. 24-25 for the relationship of the equinoxes to the seasons.)

Keep in mind that the celestial sphere is indefinitely large. All that concerns us are the *directions* of celestial objects from us. For most purposes, it makes no difference whether the center of the celestial sphere is assumed to be at the center of Earth or at our location. Remember also that the illustration here and on other pages shows the celestial sphere as seen from the "outside," while we on Earth are looking out from inside. The celestial sphere is the "screen" on which the drama of the sky is played.

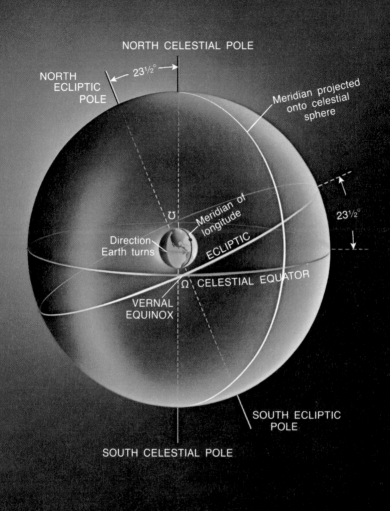

CELESTIAL COORDINATES

To specify an object's location on the celestial sphere, *equatorial coordinates* are used. These refer to the celestial equator.

The north-south coordinate (similar to latitude) is called *declination* (Dec., or δ, the lower-case Greek letter "delta"), measured in degrees north (+) and south (−) of the celestial equator, which is at declination 0°. The celestial poles are at + 90° and − 90°.

The east-west coordinate is called *right ascension* (R.A., or α, the lower-case Greek letter "alpha"). Its zero-point is the vernal equinox. Right ascension is expressed sometimes in degrees, but more commonly and usefully in hours, minutes, and seconds. Since Earth turns 360° in about 24 hours, one hour (1h) of right ascension (or time) is 15° of arc, one minute of time (1m) is 15 minutes of arc (written 15′), and one second of time (1s) is 15 seconds of arc (15″).

A degree of declination always represents the same distance on the celestial sphere, but an hour of right ascension represents a shorter distance on the celestial sphere as we move farther away from the celestial equator. In other terms, 1° of arc in R.A. always equals 4m of time, but represents a shorter distance on the celestial sphere as declination increases. Thus an object at Dec. 0° (that is, on the celestial equator) seen in a fixed telescope with a field of 1° will cross the field in 4 minutes, but at Dec. 45° the object would be in the field about 6 minutes.

The vernal equinox is right ascension 0h, and the opposite point, the autumnal equinox, is 12h, from which we go on to 24h, which is the same as 0h. Right ascension increases counterclockwise—that is, toward the east—around the north celestial pole. As observers we are inside the sphere; hence the sky pattern is reversed compared to the illustration, which shows the pattern from outside.

In the illustration are two stars: #1 and #2. Their positions can be specified thus:

Star #1: α = 2h45m00s, δ = + 62°26′56″.
Star #2: α = 22h40m55s, δ = − 36°44′05″.

Celestial longitude is similar to right ascension, except that it is measured along the ecliptic rather than along the celestial equator. *Celestial latitude* is the angular distance of an object from the ecliptic. These coordinates are useful for specifying the positions of planets, since planets are always close to the ecliptic.

Because it is curved, when the celestial sphere is represented on a flat page, distortions occur—areas get squeezed or stretched out of shape. The bottom illustration shows a flat map with the celestial equator and the ecliptic marked, along with one or two stars. Study carefully the two methods of showing the same thing. Both methods are useful in depicting the sky and will be used later in this book.

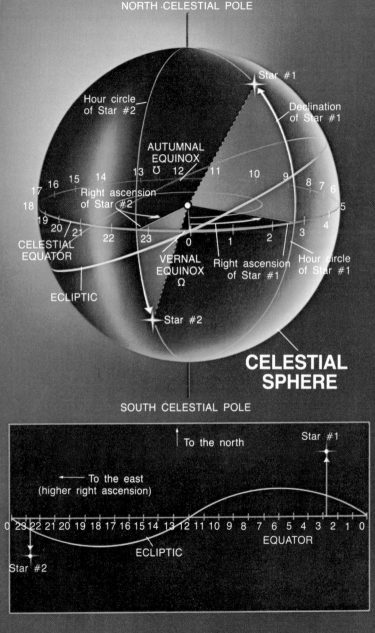

NORTH CELESTIAL POLE

Star #1

Hour circle of Star #2

Declination of Star #1

AUTUMNAL EQUINOX ♍ 12

Right ascension of Star #2

CELESTIAL EQUATOR

ECLIPTIC

VERNAL EQUINOX ♈

Right ascension of Star #1

Hour circle of Star #1

Star #2

CELESTIAL SPHERE

SOUTH CELESTIAL POLE

To the north

Star #1

To the east (higher right ascension)

ECLIPTIC

EQUATOR

Star #2

TIME AND THE SKY

Most timekeeping and calendrical schemes depend on celestial motions. A *day* is the time it takes for one complete rotation of Earth, a *month* for a complete cycle of Moon phases, a *year* for one circuit of Earth's orbit.

Unlike common clocks, the "face" of the celestial clock moves while the "hand" stands still. For observers in the northern hemisphere, the center of the clock face is the NCP (north celestial pole), marked approximately by the star Polaris (p. 76). Since the plane of an observer's horizon is tangent to Earth's surface at his location, the altitude of the NCP from his location is the same angle as his latitude. At latitude $+40°$, the NCP is 40° above the north point on the horizon. All stars within that many degrees of the NCP will always be above the horizon. Similarly, all stars within that many degrees of the SCP will never rise. Since the celestial equator is perpendicular to the axis between the celestial poles, for any observer anywhere it will cross the sky and intersect the horizon at the due-east and due-west points, and its angle with the horizon will be 90° minus the observer's latitude.

For any observer, the "hand" on the clock is an imaginary line in the sky called the *celestial meridian,* or just *meridian.* It is the observer's meridian of longitude extended out to the celestial sphere. It runs from the due-north point, through his zenith, to the due-south point, and underneath Earth back to north. The meridian for one location is different from the meridian for any other location to the east or west. An observer due north or south of you shares the same meridian, but for that observer the NCP is at a different altitude.

An imaginary circle drawn on the celestial sphere through the celestial poles and a celestial object is the *hour circle* of that object. The angle, or arc, between the celestial meridian and the hour circle of the object is the *hour angle.* Since the sky moves westward, the hour angle increases westward, from 0h on the meridian through 24h. It decreases eastward. Thus an object in the eastern sky has an eastern, or negative, hour angle. The hour angle is the measure of how long ago (or before) the object was (or will be) on the meridian.

In the illustration, the Sun's hour angle is about 2h, since it is about 30° west of the meridian. Our time of day is defined by the hour angle of the Sun, but since we want a new day to begin at midnight, when the Sun is crossing the invisible lower meridian, we use this definition: Local Apparent Time = Hour Angle of the Sun + 12h. The time is *apparent* because it is according to where the Sun appears, and *local* because it depends on the observer's location. A person east of you has a later local apparent time (a later time shown on a sundial), because the Sun crosses his meridian *before* it crosses yours. A person west of you has an earlier local apparent time because the Sun will cross his meridian *after* it crosses yours.

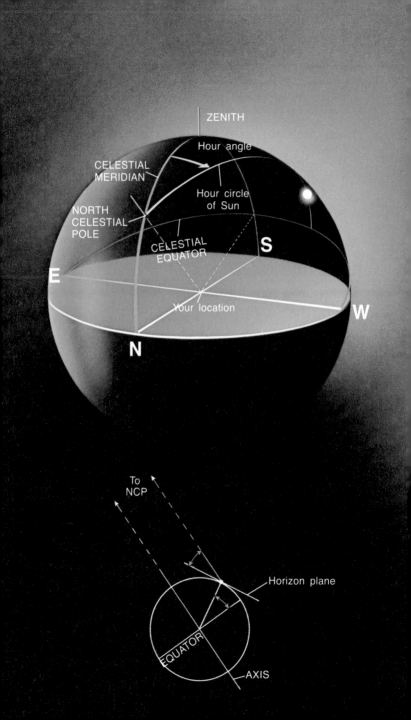

TIME ZONES

Before the days of fast communication and travel, differences in local apparent time between places a few tens of miles east or west of one another did not matter. Today it is necessary to standardize local times so that a person will not encounter different times in each different town he visits.

Standardization has been done by establishing 24 *time zones* around the world, each 15° wide. Within a zone, every location has the same *standard time*. Zones are centered on *standard meridians* at longitudes 0°, ±15°, ±30°, and so on, thus differing by 0h, ±1h, ±2h, and so on, from the time at Greenwich, England, on the prime meridian (see p. 8).

For political and practical reasons, some boundaries between zones are irregular; there are also areas where the time in the zone is a half hour more or less than in the adjoining zones. Time-zone boundaries generally appear in atlases, almanacs, and road maps, but may change.

Some areas adopt a standard time one hour later on the clock than the geographic zone they are in. At certain times of the year, some other regions, including most of the United States, adopt *daylight time*, which sets the local time one hour later on the clock than standard time, thus making daylight last farther into the evening. In the United States, daylight time is usually in effect from the first Sunday in April until the last Sunday in October. The mnemonic for adjusting the clock is "Spring ahead, Fall back."

Time zones covering the contiguous United States are: Eastern, at 75° or +5h; Central, at 90° or +6h; Mountain, at 105° or +7h; and Pacific, at 120° or +8h. Alaska, Hawaii, and United States possessions have different times. The number of hours tells you how many hours your clock is earlier than *Greenwich Mean Time*, or GMT. Astronomical almanacs often give times of events in GMT, which is also the same as *Universal Time*, or U.T., and is sometimes called *Zulu* or *Z-time*. You can find the time difference between two zones by subtracting the lower number from the higher. Thus, Philadelphia (Zone +5h) and Fresno (Zone +8h) are three hours apart, and at any given moment Fresno's clock shows an earlier time.

Unless you are located exactly on a standard meridian, your local time (more properly, local mean time; see p. 20) differs from your zone time by 4m for every degree you are off the standard meridian. Your local time is later than that shown on the standard clock if you are east of the standard meridian. For example, Seattle is 2¼° west of longitude 120°, its local time is thus 9m (2¼ times 4m) earlier than the Pacific Standard Time.

Tables of risings and settings of Sun, Moon, and planets are usually given in local mean time. You can find the exact time of the event at your location in terms of the standard time for your zone by using the method described above.

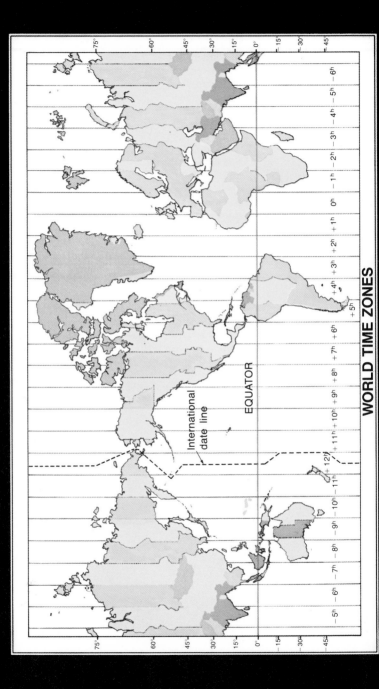

WORLD TIME ZONES

SUNDIALS AND THE EQUATION OF TIME

The origin of the sundial is lost in the past. The principle is simple and elegant: the shadow of a stick can be used to tell time. The "stick" on a sundial is called the *gnomon*, or *style*. Sundials are marvelous combinations of science and art, as the dials on the facing page show. In older cities of the world sundials on buildings or monuments are common. As late as the 19th century, some people carried pocket sundials (and even moondials!).

Sundials have ranged in size from coin-size pocket instruments to a sundial in Jaipur, India, with a gnomon 56 feet high, which you can walk up. Materials have ranged from stone and iron to gold and lucite. The earliest sundials were made around 3500 B.C. The oldest one still extant is an Egyptian shadow stone from the 8th century B.C.

Originally the daylight period was divided into 12 hours. Since the length of day varied during the year (see p. 24), the length of the hours varied also. The nighttime likewise was divided into 12 hours; these also varied with the season and differed from the daylight hours. An Arab astronomer in the 13th century invented equal hours for astronomical purposes, and this caught on when mechanical clocks were invented somewhat later.

Most dials are designed to be used in one location, examples being the vertical dial, horizontal dial, and armillary dial. The gnomon is fixed and is inclined so as to be parallel to Earth's axis. The upper end points to the north celestial pole, if the instrument is for the northern hemisphere. As the Sun crosses the sky, the shadow of the gnomon on a horizontal dial moves from west through north to east—that is, clockwise. From their beginnings, mechanical clocks have been designed so that the clock hands imitate the motion of a gnomon's shadow. Some sundials have adjustable gnomons for use at different locations and often include tables of corrections for cities the traveler is likely to visit.

The time shown by a sundial is called local apparent time (LAT). It is *local* because it depends on the dial's location; it is *apparent* because the position of the gnomon's shadow depends on the position of the Sun as it appears in the sky.

The Sun's apparent motion across the sky is not uniform all year. The Sun makes one complete trip around the sky in about 365 days. Thus, its mean motion is 360°/365 per day, or slightly less than 1° per day. Because the Sun moves along the ecliptic, not along the celestial equator (see p. 12), some time is spent moving slightly north or south as well as eastward (see top illustration, p. 23). Thus during some periods the Sun appears to move eastward faster than during others. This effect alone, due to the obliquity of the ecliptic, would cause the real Sun to get as much as 9m out of step with a uniformly moving Sun, called the *mean Sun*. Time measured by the mean Sun is *Local Mean Time* (LMT).

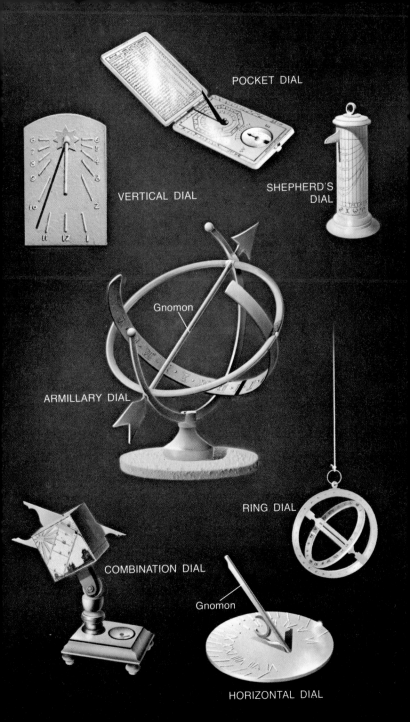

POCKET DIAL

VERTICAL DIAL

SHEPHERD'S DIAL

Gnomon

ARMILLARY DIAL

RING DIAL

COMBINATION DIAL

Gnomon

HORIZONTAL DIAL

The orbit of Earth about the Sun is not a circle but an ellipse: it is eccentric. As Kepler's laws describe (see p. 246), this means the motion of the Sun along the ecliptic is faster when Sun and Earth are relatively close (around January) and slower when they are relatively distant (around July). This effect alone would make the mean Sun run fast and slow by as much as 10^m.

These two factors, obliquity and eccentricity, are not in phase: their maximums and minimums do not coincide. Their individual effects are shown in the lower diagram on p. 23. The effect of obliquity is the dashed yellow line; the effect of eccentricity of the ecliptic is the dotted purple line. The sum of the two effects, the total amount by which the true Sun is fast or slow compared to the mean Sun, is called the *Equation of Time*, or EOT, shown by the solid white line.

The relation between the three kinds of solar time is:

$$LMT = LAT - EOT.$$

Sometimes the value of EOT, with sign reversed (that is, plus becomes minus, and vice versa) is called *correction to sundial*. It is added to sundial time (LAT) to get LMT. (To get your local standard time, correct for your offset from the standard meridian; see p. 18).

The graph on p. 23 is only approximate. It is accurate to within about a minute for any year, but for precision in any particular year, you must consult an almanac.

The fact that the Sun runs fast and slow compared with clock time means that it sometimes reaches your meridian before noon, and sometimes after noon, as measured by your clock corrected for your offset from the standard meridian. If you could mark the Sun's location in the sky each day just when your watch reads 12:00, over a year the pattern of marks in the sky would look like a stretched-out figure 8.

The position of the gnomon's shadow depends on the Sun's position in the sky. Therefore, if every day at noon (standard time) by your watch you marked the location of the tip of the shadow, over a year's time you would draw a similar figure 8. This is called an *analemma*. Such figures, often seen on world globes, are used to correct for the equation of time, since that is just what an analemma is: a folded-over graph of the EOT. Note that while the value of the EOT does not depend on your location, but rather on the time of year, the exact shape of the analemma will depend on your latitude and your offset from the standard meridian.

If you try to draw an analemma, use standard time even when you are officially on daylight time. For more details about time and about how to construct a sundial, see the Bibliography.

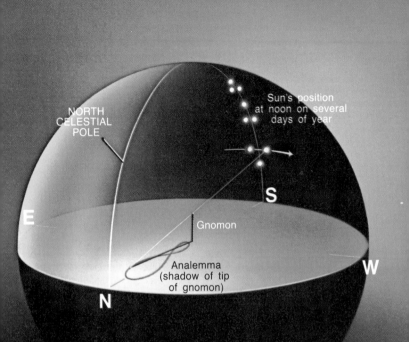

NORTH CELESTIAL POLE

Sun's position at noon on several days of year

E

S

Gnomon

Analemma (shadow of tip of gnomon)

N

W

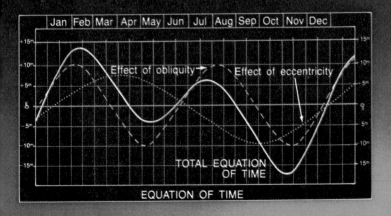

| Jan | Feb | Mar | Apr | May | Jun | Jul | Aug | Sep | Oct | Nov | Dec |

+15ᵐ

+10ᵐ

+5ᵐ

Effect of obliquity → Effect of eccentricity

−5ᵐ

−10ᵐ

−15ᵐ

TOTAL EQUATION OF TIME

EQUATION OF TIME

THE SEASONS

As Earth goes around the Sun, the Sun and planets seem to move against the background of stars. The "band" of sky through which they move is the *zodiac*, and the Sun's path in this band is the *ecliptic*. Since the ecliptic is inclined toward the celestial equator by 23½°, during the year the Sun changes its declination. It is 0° at the *vernal equinox*, about March 21; +23½° at the *summer solstice*, about June 22; 0° again at the *autumnal equinox*, about September 22; and −23½° at the *winter solstice*, about December 22. "Equinox" and "solstice" refer to both points in the sky and points in time.

The celestial equator intersects the horizon at the exact east and west points. Any celestial object with a positive declination will rise north of east and set north of west. Conversely, any object with a negative declination will rise south of due-east and set south of due-west. Since exactly half, or 12h, of the celestial equator is above the horizon, an object north of the celestial equator will spend more than 12h a day above the horizon in the northern hemisphere, and less than 12h in the southern hemisphere.

As the Sun moves north each year in the northern hemisphere, the length of daytime (during which the Sun is above the horizon) increases, the Sun's path across the sky is higher, and summer comes. The weather grows warmer because the Sun is heating that hemisphere of the earth longer each day, less heat is radiated away by the earth at night (which is shorter), and the Sun's rays with their higher angle are striking the earth more directly (more concentratedly). As the Sun moves south again, the changes are reversed: summer yields to fall, and winter comes as the Sun reaches far south. As the Sun moves north again, winter turns to spring, and spring to summer once more. The warmest weather does not coincide with the highest Sun, but comes later, in midsummer, because it takes time for Earth, oceans, and atmosphere to heat up. A similar lag occurs in winter.

In Earth's southern hemisphere, the seasons are the reverse of those in the northern hemisphere, summer occurring during the north's winter, and spring and fall during the north's fall and spring.

On p. 25 are shown lengths of day and night for latitudes 0°, +40°, and +70°. The red line shows the situation at the vernal equinox. Often it is said that at the time of an equinox, day and night are equal. Actually, equality occurs a few days before the vernal equinox and a few days after the autumnal equinox. One reason is that the Sun is not a point of light but a disk about ½° wide; thus sunrise and sunset have duration. The other reason is that refraction of solar rays by the atmosphere (see p. 40) lengthens daylight. Similarly, the solstices do not correspond with the dates of earliest and latest sunrise or sunset.

SEASONS

AQUARIUS
CAPRICORNUS
SAGITTARIUS
To background constellations
PISCES
EARTH'S ORBIT
SCORPIUS
LIBRA
ARIES
VIRGO
TAURUS
GEMINI
CANCER
LEO
CELESTIAL SPHERE

LENGTH OF DAY

At equator	MONTH	At latitude 40°N.	MONTH	At latitude 70°N.
	D			D
	N			N
	O			O
	S			S
	A			A
	J			J
	J			J
	M			M
	A			A
	M			M
	F			F
	J			J

PARALLAX AND UNITS OF DISTANCE

Earth's movement around the Sun produces a slight yearly periodic shift, called *parallax*, in observed positions of nearer stars. The parallax is one half of the yearly angular shift in position of the star. The size and shape of the star's apparent path against the background of more distant stars vary with the star's distance and its position relative to the ecliptic plane. In the top illustration Star #1, being closest to Earth, has the greatest shift. (Note: The drawing is not to scale.) During a year the star seems to move back and forth along the ecliptic. Star #3 is most distant and thus its shift is smallest. Since it lies at the ecliptic pole (or would if there were no parallax), over a year it seems to draw a small circle around the ecliptic pole. Star #2 is intermediate in distance, somewhat between ecliptic and pole; hence its shift is intermediate.

Astronomers use a unit of distance called the *parsec*, the distance at which a celestial object has a parallax of one second of arc. The more distant the star, the smaller the parallax. The nearest star, Proxima Centauri, has a parallax of only about 0.62 arc second.

Another astronomical distance unit is the *light-year*, or distance traveled by light in a vacuum in one year at 186,282 miles per second. One light-year equals 0.3069 parsec, or 5.88 million million miles. The *astronomical unit* (a.u. or A.U.), or mean distance between Earth and Sun (92,955,806 miles), is another standard for measurements. A parsec equals 206,265 a.u.

PRECESSION

Earth's axis is tilted $23\frac{1}{2}°$ from the vertical to the ecliptic plane. Both the Sun, always on the ecliptic, and the Moon, always within 5° of it, are continuously "trying" to make Earth upright by their gravitational pull on Earth's small equatorial bulge (much exaggerated here). This pull gives Earth's axis a wobble instead of causing it to become more nearly vertical with respect to the ecliptic. This wobble is called *precession*. One wobble, taking about 26,000 years, causes each end of the axis to describe a small circle in space.

Because of precession, star positions as we see them change slowly. Even the position of Polaris, our north star, is slowly changing. Polaris will draw closer to the north celestial pole until about A.D. 2105, then pull away again. When the Egyptian pyramids were built, the star Thuban, in the constellation Draco, was the north star.

Because the pulls of Sun and Moon are not uniform, the precessional wobble is not smooth. There is a slight "nodding" of the axis, called *nutation*, represented (with exaggeration) in the figure by waves superimposed on the precessional circle. Nutation can be ignored by amateurs, but not by professionals. After decades, catalog positions of stars must be corrected; hence star coordinates are dated.

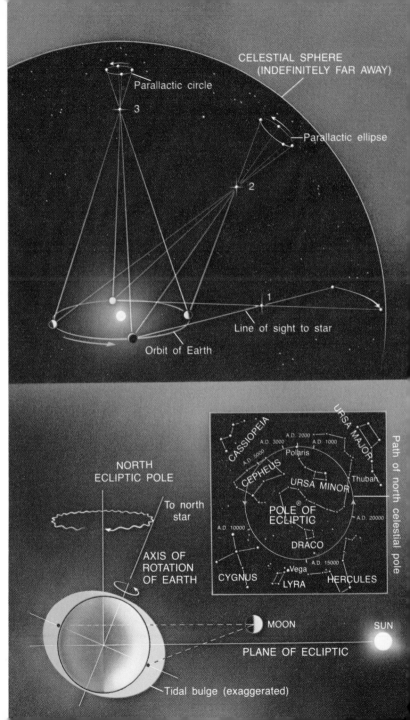

CELESTIAL SPHERE (INDEFINITELY FAR AWAY)

Parallactic circle

3

Parallactic ellipse

2

1

Line of sight to star

Orbit of Earth

NORTH ECLIPTIC POLE

To north star

AXIS OF ROTATION OF EARTH

CASSIOPEIA

URSA MAJOR

A.D. 3000   A.D. 2000

A.D. 1000

A.D. 5000

CEPHEUS

Polaris

Thuban

URSA MINOR

POLE OF ECLIPTIC

A.D. 20000

DRACO

Path of north celestial pole

A.D. 10000

A.D. 15000

CYGNUS

Vega

LYRA   HERCULES

MOON

SUN

PLANE OF ECLIPTIC

Tidal bulge (exaggerated)

BASIC OPTICS

The operative components of astronomical instruments such as telescopes, binoculars, and cameras are lenses, mirrors, and prisms. Some knowledge of how they work, and of their limitations, is necessary for an understanding of what equipment to buy and how to use it.

A lens or mirror is *positive* if it causes incoming light rays to converge. The distance from the center of the lens or mirror to the point *(focal point,* or *focus)* where the parallel rays focus is its *focal length.*

Since light can pass through a lens in either direction, light from a source at the focal point passing through the lens will converge into parallel rays. In the top illustration, either arrowhead can be thought of as the object, the source of light; the other is the *image.*

Negative lenses or mirrors cause light to diverge, as in the illustrations second from the top, and at bottom center. A negative lens is concave; a negative mirror is convex. They are said to have negative focal lengths.

A concave mirror is positive, reflecting parallel light rays to a focus. But a telescope mirror curved like the surface of a sphere would not bring parallel rays to a single focus, as shown in the illustration at bottom left; *spherical aberration* would result. Hence, the curve must be altered slightly to a paraboloidal form, so that all rays come to a single focus (bottom right).

A mirror reflects light of every color in the same way, but a lens is like a stack of prisms. Each prism refracts light of different colors differently—blue light, for example, more than red light. A *spectrum* is produced when white light passes through. Likewise, a lens brings rays of different colors to foci at different distances. A telescope with only a single lens would make a white star look like a series of concentric colored rings—an effect called *chromatic aberration.* A good telescope has a composite *achromatic lens,* consisting of two or more lenses of different strengths and different kinds of glass that compensate and make the rays of any two colors focus at the same point.

The quality of image produced by an optical instrument depends on the quality of the components—how carefully they are made, how accurately they are aligned with one another. Variations in components account for the wide range in quality, and hence price, of optical devices.

Lenses and mirrors demand care in use as well as occasional maintenance. Smears and scratches, even if slight, impair the quality of the image. Avoid touching the surface with the fingers, and clean it only with a camel's-hair brush or lens-cleaning tissue (with a rotary motion) or otherwise follow the manufacturer's instructions. Some telescopes, especially reflectors, periodically need realignment, which can usually be done by the amateur with the directions provided.

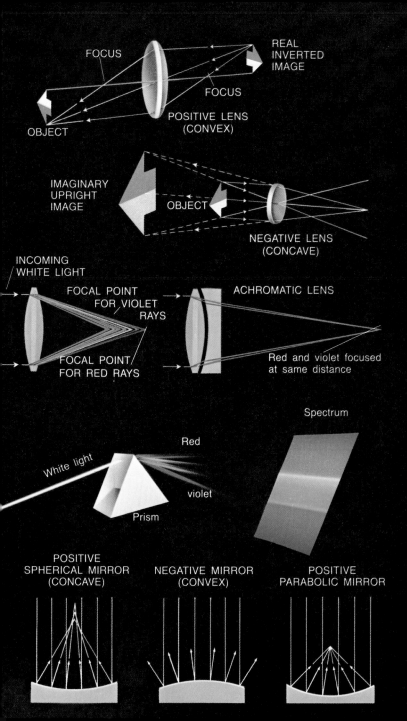

FOCUS

FOCUS

REAL INVERTED IMAGE

OBJECT

POSITIVE LENS (CONVEX)

IMAGINARY UPRIGHT IMAGE

OBJECT

NEGATIVE LENS (CONCAVE)

INCOMING WHITE LIGHT

FOCAL POINT FOR VIOLET RAYS

FOCAL POINT FOR RED RAYS

ACHROMATIC LENS

Red and violet focused at same distance

Spectrum

Red

White light

violet

Prism

POSITIVE SPHERICAL MIRROR (CONCAVE)

NEGATIVE MIRROR (CONVEX)

POSITIVE PARABOLIC MIRROR

TELESCOPES

Telescopes are divided into three main classes: *refractors*, which use lenses to gather light; *reflectors*, which use mirrors; and *catadioptric* systems, which use a combination of lens and mirror.

The main light-gathering component of any telescope is the *objective*. The greater its area, the greater the amount of light received. This increases as the square of the objective's diameter. The objective does not magnify; the task is for supplementary lenses. But adequate light grasp is needed to detect faint objects and render good image detail. A 3-inch lens gathers about 100 times as much light as the eye, which is why you must *never* view the Sun through a telescope without precautions (p. 206).

An important tool in evaluating an objective is its *f-ratio*—the ratio of its focal length (F) to its diameter (D). Thus, where $F = 48$ inches and $D = 6$ inches, $F/D = 8$, the f-ratio, sometimes written f/8. An f-ratio of 3 to 6 is useful for wide fields of view and faint objects, but image size is small. For larger images, showing more detail on Moon or planets, an f-ratio of 10 to 20 or more is preferred. A telescope with an objective in f/8 to f/10 range is practical for amateurs because it represents a compromise between wide field and high magnification.

To view an image formed by the objective, a lens called an *ocular*, or *eyepiece*, is placed near the focus. Eyepieces of different focal lengths provide different magnifications. *Magnification* (M) equals the focal length of an objective (F) divided by the focal length of the eyepiece (f). The use of magnifications less than about 3 or more than 50 times the diameter in inches of the objective usually will give poor results.

In a reflector the objective is a mirror, called the *primary*. Because it reflects light (instead of transmitting it), another mirror, a small *secondary*, usually is introduced into the reflected beam to direct it to a better position for viewing. The amount of incoming light blocked by the smaller mirror is usually negligible. Two commonly used primary-secondary systems are shown; there are many others.

Catadioptric telescopes, such as the Schmidt and Maksutov systems, have a mirror with a spherical curve, so that a lens is needed to correct for aberrations. These telescopes typically have a wide field, small f-ratio, and longer focal length within a short space.

Resolving power is the ability of a telescope to distinguish between extremely close objects. The theoretical resolving power of an objective with D measured in inches is $4.54/D$ arc seconds; thus a good 6-inch telescope will "split" stars 0".76 apart.

To aim a telescope at a celestial object, a *finder* is used: a small telescope with a wide field, mounted on and aligned with the main instrument. With its wide field, the finder is easy to aim; and when it is aimed properly, so is the main instrument.

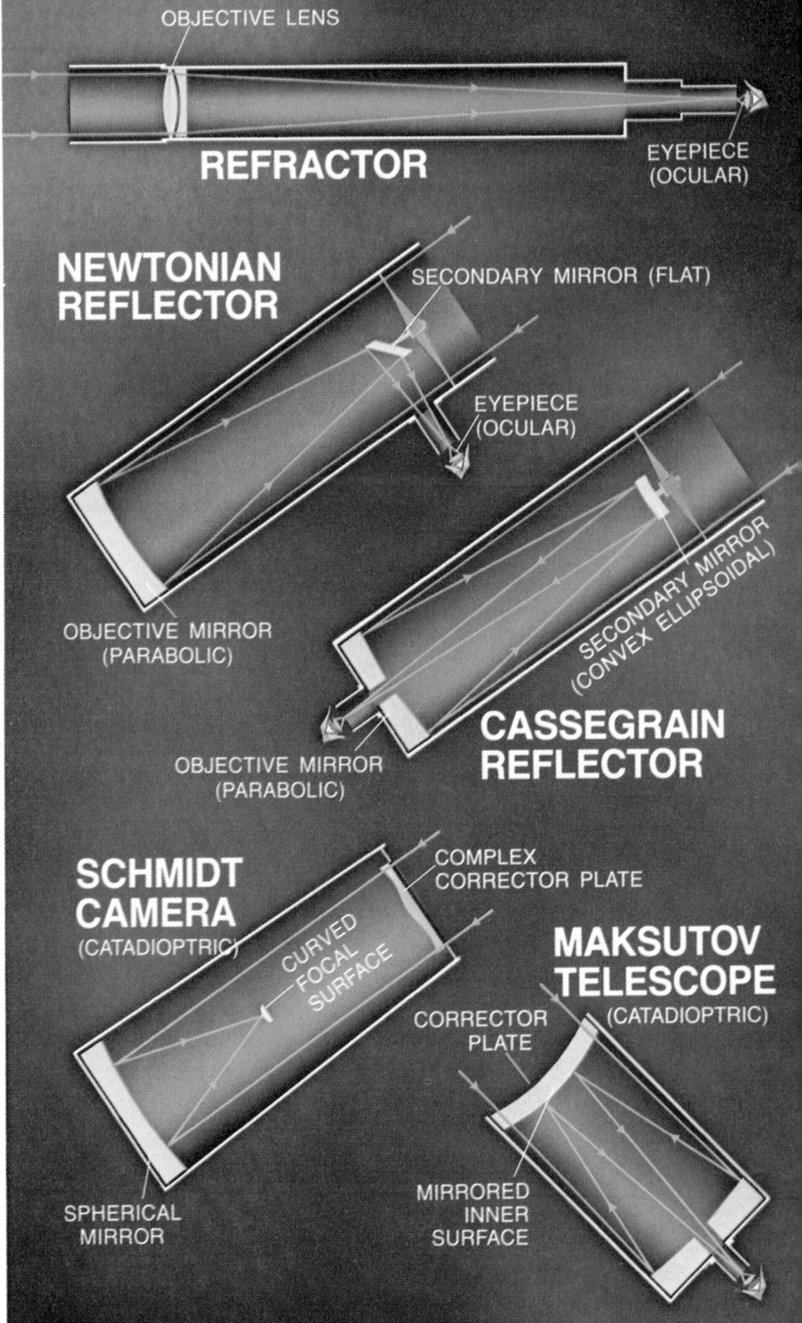

OBJECTIVE LENS

REFRACTOR

EYEPIECE (OCULAR)

NEWTONIAN REFLECTOR

SECONDARY MIRROR (FLAT)

EYEPIECE (OCULAR)

OBJECTIVE MIRROR (PARABOLIC)

SECONDARY MIRROR (CONVEX ELLIPSOIDAL)

CASSEGRAIN REFLECTOR

OBJECTIVE MIRROR (PARABOLIC)

SCHMIDT CAMERA (CATADIOPTRIC)

COMPLEX CORRECTOR PLATE

CURVED FOCAL SURFACE

MAKSUTOV TELESCOPE (CATADIOPTRIC)

CORRECTOR PLATE

SPHERICAL MIRROR

MIRRORED INNER SURFACE

TELESCOPE MOUNTINGS

The purpose of a telescope mounting is to hold the telescope, keep it steady, and enable the observer to keep it pointed at a celestial object as Earth rotates. The importance of quality can hardly be exaggerated. A poor mounting is hard to adjust and may shiver at a touch, or in the wind, or from any ground vibration, spoiling observations. The two basic types of mountings are the altazimuth and the equatorial.

ALTAZIMUTH MOUNTING: This mounting, similar to a camera tripod, is illustrated at the bottom of the facing page. Two axes of rotation, one vertical and one horizontal, allow the telescope to move in altitude and azimuth. Although well-suited for terrestrial observations, this mounting is not convenient for astronomical viewing because sky objects continuously change in both altitude and azimuth, as depicted on the opposite page by the red arrow. Here the star, in the southeastern sky, is moving upward and southward. Thus the mounting must rotate about both axes simultaneously at continuously varying rates to keep the star in view. Such mountings are found usually on cheap astronomical telescopes—or, occasionally, on better ones that have a very wide field of view.

EQUATORIAL MOUNTING: Designed for serious stargazing, this mounting has one axis of rotation—the right-ascension axis, or *polar axis*—parallel to Earth's rotation axis. The other axis of rotation, at right angles to the polar axis, is the *declination axis*. Rotation about this axis keeps the telescope aimed along a celestial meridian; that is, it changes only the declination of the field of view (blue arrows).

To guide the observer in adjusting the axes, the mounting is usually equipped with *setting circles,* one on the polar axis and the other on the declination axis. Once the declination axis is properly set for a given star, rotation about the right-ascension axis will keep the star in view, because its declination is not changing and its hour angle is changing at a constant rate (purple arrow).

The telescope can be moved by hand or—far more convenient—by an electric clock-driven motor, geared to rotate the telescope at the proper rate. The rate is based on the *sidereal day,* which is the time that elapses between two successive transits of the first point of Aries over the upper meridian. The sidereal day is about 23h 56m 4s, being shorter than the solar day (averaging about 24 hours) because the revolution of Earth around the Sun causes the transit to occur about 4m earlier each night.

Exact tracking with an equatorial telescope requires that the mounting be adjusted for the observer's latitude and that the assembly be placed level with the polar axis pointed at the celestial pole. For many amateur purposes, pointing at the north star by eye is sufficient. For photographic time exposures, precise orientation is necessary.

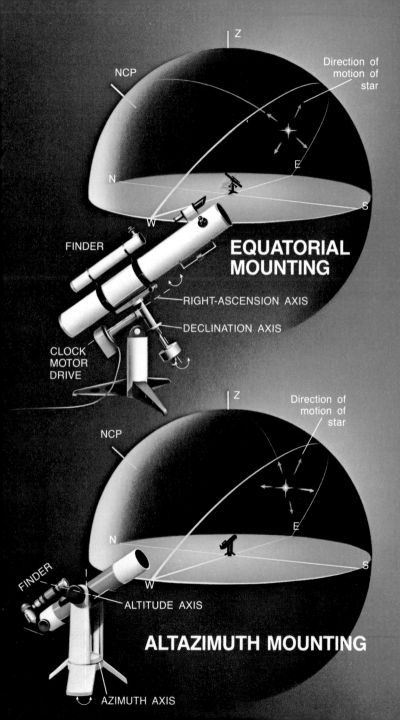

Z

NCP

Direction of
motion of
star

N

E

W

S

EQUATORIAL
MOUNTING

FINDER

RIGHT-ASCENSION AXIS

DECLINATION AXIS

CLOCK
MOTOR
DRIVE

Z

NCP

Direction of
motion of
star

N

E

W

S

FINDER

ALTITUDE AXIS

ALTAZIMUTH MOUNTING

AZIMUTH AXIS

ASTROPHOTOGRAPHY

For photographs of the sky you need an adjustable camera, capable of taking time exposures, and a steady mounting. A 35mm camera, preferably a single-lens reflex, is versatile and convenient. In general, a relatively fast film—ASA 200 or faster—is required.

Films, particularly color films, may show the phenomenon called *reciprocity failure*. This occurs when a film designed for everyday use is exposed for a period longer than it was designed for. Beyond a certain point, increasing the exposure time adds nothing to the image. This limitation may be important when one is making long exposures of relatively faint objects.

Color film tends to be slower and grainier than black-and-white. However, it does show celestial objects in color—color that may be very interesting, though not necessarily much like what the eye sees. Experts sometimes chill their film to dry-ice temperatures for better results.

In all astrophotography, use a cable release to avoid vibrating the camera when tripping the shutter. If necessary, shield the camera from the wind and from nearby lights, such as from houses or passing cars.

Following is a brief introduction to astrophotography. For the finer points, consult an advanced book on the subject.

UNGUIDED PHOTOGRAPHY: In this kind of photography a camera is mounted on a sturdy support (such as a tripod or even a beanbag), focused at infinity, and pointed at the sky. If the film is exposed for a few minutes or more, all the celestial objects bright enough to register will make trails on the film, because of Earth's rotation. If the camera is pointed north, the trails will be arcs centered on Polaris. If the camera is pointed in some other direction, the trails will be longer and more broadly curved. Trail length depends on duration of exposure *and* distance of the object from the celestial equator—the nearer to the equator, the longer the trail. However, each trail will represent the same proportion of a full circle.

With a typical 50mm lens, star images trail little with an exposure of 7 sec. or less. Longer exposure is possible with a wide-angle lens, shorter exposure with a telephoto. The faster the film and the clearer the sky, the greater the number of stars that can be registered by an exposure of given length, unless reciprocity failure occurs.

Unguided photography makes possible a wide variety of pictures—star and meteor trails, comets, Moon and Sun (see p. 206!), the brighter constellations (usually requiring a wide-angle lens), bright star clusters such as the Pleiades, and—among the planets—Venus at least.

GUIDED PHOTOGRAPHY: If your camera can be kept pointed at a celestial object while Earth rotates, long exposures can be made without causing images to trail, and relatively faint objects can be photo-

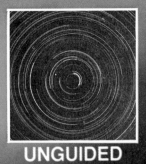
UNGUIDED

GUIDED
WIDE FIELD

GUIDED

THROUGH
TELESCOPE

graphed. The camera can be put on an equatorial mounting, preferably with a clock drive and a finder. The drive compensates for Earth's rotation, keeping the camera pointed at the object. While the drive is working, the photographer looks through the finder and, if necessary, makes small guiding adjustments to keep the celestial object centered in the field. If the camera is mounted on an equatorial telescope, guiding can be done as you look through the telescope or finderscope.

THROUGH-THE-TELESCOPE PHOTOGRAPHY: Here the telescope's mirror or lens can be the objective of the camera. Telescopes usually have much longer focal lengths than even telephoto lenses; hence the field of view is smaller, but magnification is greater and a clock drive therefore is almost imperative.

At prime focus: The camera without its lens is mounted on the telescope with a special adapter. The telescope's objective focuses light directly on the film in the camera. Focusing is done by moving the camera backward or forward on the adapter until the image is sharp as seen through the camera's viewer. If the camera is not a single-lens reflex, focusing is done by inserting a ground glass in the camera at the film position and then fixing the camera at the point where the image is sharp. At prime focus, the image of the Moon produced by an objective of 50 in. focal length is about ½ in. wide.

Barlow-lens projection: A "Barlow" is a negative lens placed in the optical path to increase the effective focal length of the objective and thus increase magnification, usually by about 3X.

Eyepiece projection: The camera is mounted on the telescope so as to receive the light from the telescope's eyepiece, which is at the prime focus. The image is relatively large. Because the light is spread widely, longer exposures are needed.

Eyepiece-plus-camera-lens: Both eyepiece and camera lens are used. Image quality depends on both. Each lens, having slight aberrations, will absorb some light as it passes through. Experiment with various eyepiece/lens combinations for best results.

ASTROPHOTOGRAPHY GUIDE: On the facing page is a chart adapted from an Eastman Kodak Company publication (see p. 267). In all cases, fast black-and-white or color film can be used. Lunar photographs require only medium or slow-speed film (ASA 50 to 200, approximately). Solar photographs (CAUTION! See p. 206) are made with very slow, fine-grain film for best resolution. Some film types can be "pushed" by the laboratory; hence exposure times can be shortened somewhat. To be sure of getting proper exposures, some astrophotographers use the "bracketing" method, taking each picture with the recommended exposure and then again with longer and shorter exposures.

Object	Type of Instrument	Mounting Type	Objective	f-ratio	Recommended Exposure
Star movement and comet trails	Camera with time-exposure capability	Rigid support	Any lens	Wide open	2 to 30 min.
Meteors	Camera with time-exposure capability	Rigid support	Wide-field lens	f/6.3 or faster	10 to 30 min.
Aurorae	Camera with time-exposure capability	Rigid support	Fast lens	f/4.5 or faster	1 sec. to 2 min.
Moon	Camera or camera with telescope	Rigid, or guided with or without drive	20mm or larger	f/4.5 or slower	$\frac{1}{125}$ sec. to 10 sec.
Stars and comets	Camera or camera with telescope	Equatorial with guiding	25mm or larger	f/6.3 or faster	10 min. to 1 hr.
Star clusters, nebulae, galaxies	Camera or camera and telescope	Equatorial with guiding and drive	25mm or larger	About f/6.3	10 min. to 1 hr.
Planets	Camera with telescope	Equatorial with drive	25mm and up; best with 150mm and up	Varies with your system	$\frac{1}{2}$ to 15 sec.
Sun Caution! See p. 206	Camera or camera with telescope	Rigid or equatorial	Neutral density filters *over* 25 to 100mm main objective	f/11 to f/32	$\frac{1}{1000}$ to $\frac{1}{30}$ sec.
Artificial satellite	Camera with time-exposure capability	Rigid	Fast lens	f/4.5 or faster	Duration of pass

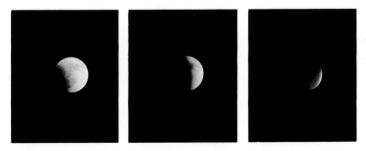

Progress of total lunar eclipse photographed with a 350mm lens and ASA 400 color film: 1/125, 1/30, and 2 sec. During totality (right) Moon remains visible, with coppery color, because some sunlight is refracted onto Moon through Earth's atmosphere.

BINOCULARS FOR SKYWATCHING

Binoculars are good for scanning wide areas of sky and viewing the Moon. They are particularly good for your first "close-up" of the heavens. The image in binoculars is upright and visible with both eyes. Fairly good binoculars are less costly than the least expensive telescope of good quality. They are a good investment, since even if interest wanes, binoculars can be used for other purposes.

Get real binoculars, not "opera glasses." The latter are merely two simple Galilean telescopes side by side. They are smaller than binoculars and provide lower magnification, a smaller field of view, and an image of poorer quality. Some are made to look like binoculars; so—to be safe, buy from a reputable source.

True binoculars have a pair of prisms in each optical path to produce an upright image and make possible a long optical path within a short instrument length. Binoculars are available with magnifications from 3x to 20x or more, and with objective lenses from a few millimeters to several centimeters in diameter. They are usually classified by a pair of numbers such as "7x50," the first number indicating the magnifying power of the instrument, the "x" meaning "times," and the second number the diameter of each objective lens measured in millimeters. A 7x50 is a good all-around type for astronomical purposes. In general, lower-powered binoculars have wider fields; higher-powered ones are heavier and harder to hold steady by hand.

The higher the power with an objective of a given diameter, the more the light from the object is spread out and the less bright is the image. Often binoculars are rated in terms of Relative Light Efficiency, or RLE, easily calculated by dividing the diameter of the objective lens in millimeters by the square of the power. For a 7x50 instrument the RLE is 50/(7x7), or about 1. A lower RLE means the image is less bright. An RLE of less than 1 is not recommended for astronomy, except for viewing the Moon. An RLE of 1 will enable you to see objects about 20 times fainter than the unaided eye can detect.

For prolonged viewing, particularly with higher-powered binoculars, some support is needed. Various clamps are available for use with a camera tripod. This arrangement is an altazimuth mounting, but because the field is large it is not a serious drawback. Some amateurs have devised equatorial binocular mounts.

Another useful accessory, available from suppliers, is a pair of small rubber cups to fit over the eyepieces of the binoculars to keep out extraneous light during viewing.

Never view the Sun through binoculars—they would act like burning glasses! Prolonged viewing of a full Moon also is inadvisable. On a clear, dark night 7x50 binoculars steadily held can reveal objects as faint as 9th magnitude, including star clusters such as the Beehive and Hyades, and even the planet Jupiter's four brightest satellites.

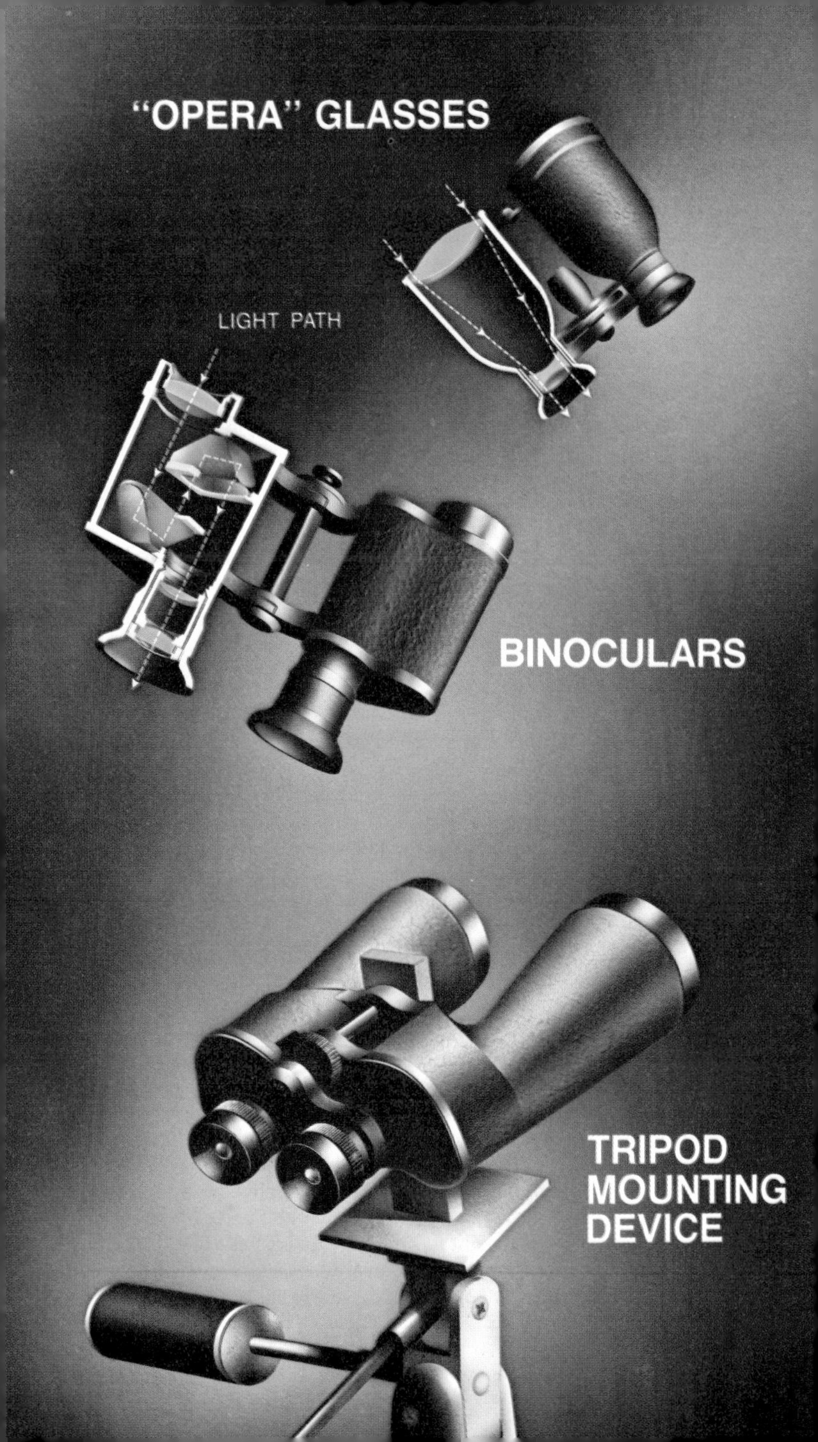

"OPERA" GLASSES

LIGHT PATH

BINOCULARS

TRIPOD
MOUNTING
DEVICE

AURORAE

The word "aurora" means "dawn," and thus is a complete misnomer for the *aurora borealis*, the northern lights, and the *aurora australis*, the southern lights. Both are produced by the glow of gas molecules high in the atmosphere—a glow caused by the solar wind, a stream of high-energy particles flung out by the Sun. Occasionally there are extra-strong gusts that reach Earth in a few days. Since the particles—mostly protons—are charged, the magnetic field of Earth deflects them toward Earth's magnetic poles. These magnetic poles do not coincide with the geographic poles. The north magnetic pole is located in northern Canada at about latitude 78.6°N., longitude 70.1°W., and the south magnetic pole at 78.5°S., 106.8°W. When the high-speed incoming particles strike molecules and atoms of oxygen, nitrogen, and other gases several hundred miles up, the molecules become excited, or raised to a higher energy state. They expel extra energy by radiating light, much as a neon bulb does.

Molecules and atoms of different gases radiate different colors. A spectroscope reveals auroral light as radiations from atoms and molecules of oxygen and nitrogen, with some radiation from hydrogen atoms also.

An aurora may take any of many forms, such as rays, arcs, curtains, or just a diffuse glow over part of the sky. Colors range from red to green to yellow. Faint aurorae look whitish because they are below the intensity needed for color detection by the eye. The various auroral forms characteristically flicker or undulate.

Most aurorae occur in the "auroral ovals": the roughly oval bands surrounding the geomagnetic poles. In the top figure here the greenish area is an oval. Aurorae occur more frequently at times of greatest sunspot activity, and it is particularly then that aurorae are seen in latitudes far from the poles. Rarely, aurorae have been seen from near the equator.

It is not possible to predict exactly when and where an aurora will be seen. Monitoring of solar "storms" can alert us to the strong possibilities of auroral activity, but is no guarantee. When aurorae do occur, they are more common and brighter near the poles. A clear, dark sky is best for observing, but not essential. Aurorae have been seen from within cities. Bright aurorae can be photographed with a stationary camera.

Times of auroral activity are times of increased activity in Earth's magnetic field and the ionosphere. Serious effects on radio communications and even on electric power transmission may occur.

Even without auroral activity, the sky has a faint light called *airglow*, produced somewhat like the aurora but much less intense and more evenly distributed. Airglow does not substantially affect observation, except for long exposures with the largest telescopes.

Path of a charged particle

Auroral oval where most aurorae occur

Line of force of Earth's magnetic field

RAYED AURORA

CURTAIN AURORA

ARC AURORA

ZODIACAL LIGHT

The material in our solar system ranges in size from our enormous Sun down to microscopic dust. The latter is responsible for two subtle phenomena that can be seen only under the excellent observing conditions of clear, dark skies.

The *zodiacal light* is a faint glow in the sky, roughly triangular, along and near the ecliptic. At its brightest and under good observing conditions, it is about as bright as portions of the Milky Way. It is the faint reflection of sunlight from dust grains, most of which lie near the plane of the ecliptic in the solar system. It is brightest at small angular distances from the Sun, and fades off as our gaze turns away from the Sun. The light looks like the glow shown in the lower illustration on the opposite page, and is best seen just after sunset and just before sunrise (the latter appearance being responsible for its older name, "false dawn").

Zodiacal light is most easily visible at seasons when the ecliptic is most nearly perpendicular to the horizon, favoring after-sunset viewing in March and April and before-dawn viewing in September and October, when viewed from mid-northern latitudes. Viewing in the tropics is even better, since the ecliptic is more nearly vertical there, as the diagrams on p. 15 show.

GEGENSCHEIN

Exactly opposite the Sun, an even fainter reflection, or backscattering, of sunlight causes the *gegenschein*, or *counterglow*. It appears as a "blob" of light about 8° long and 6° wide, only $\frac{1}{15}$ to $\frac{1}{30}$ as bright as the zodiacal light. Very careful observations reveal that it is a part of the zodiacal light, but the connection is too faint to be seen by the eye. The gegenschein is visible only under exceptionally clear skies, and is sometimes mistaken for one of the fainter "star clouds" of the Milky Way. Unlike those star clouds, it does not resolve itself into stars when examined with a telescope.

Much of the meteoric dust responsible for these phenomena is believed to come from comets. The particles are non-metallic, and about a hundred-thousandth of a centimeter in size. Every time a comet passes close to the Sun, some of its ices sublimate and release trapped dust. This then disperses through the solar system, particularly the inner region. Some of it is concentrated near Earth. Trailing Earth as it moves around the Sun may be a cloud of dust about a million miles long, ending where the attractive force of Earth's gravity and the repulsive force of the solar wind balance. A number of our interplanetary spacecraft have investigated the distribution of dust in the solar system, but much remains to be learned.

ZODIACAL LIGHT

GEGENSCHEIN
(from direction opposite Sun)

THE MAGNITUDE SCALE

Celestial objects appear in a vast range of brightnesses. The midday Sun is 16 trillion times brighter than the faintest stars just visible to the dark-adapted eye. The faintest stars photographable with our largest telescopes appear only a hundred-billion-billionth as bright to us as our Sun. Such numbers are hardly understandable and some other means of expressing astronomical brightnesses was needed. Hence the *magnitude scale*: a logarithmic scale in which a *difference* of exactly five magnitudes corresponds to a *factor* in intensity of 100 times. Thus one magnitude, written 1m, corresponds to a factor of the fifth root of 100, or 2.512, so that:

> 1m = a factor of 2.512, or about 2½
> 2m = a factor of 6.310, or about 6¼
> 3m = a factor of 15.840, or about 16
> 4m = a factor of 39.811, or about 40
> 5m = a factor of 100, exactly.

For example: Suppose two stars differ in magnitude by 13m. To find the difference in brightness in arithmetic terms, first break 13m down into as many 5m intervals as possible; thus, 13m = 5m + 5m + 3m. Replace each 5m by the number 100 and each + by ×. From the table above, you can find that 3m corresponds to another factor of 16. Thus, 13m = 100 × 100 × 16 = 160,000 (times as bright). For the greatest possible exactness, magnitudes may be given to one or more decimal places; e.g. 1.80 for the star Algenib.

The magnitude scale is set up in such a way that light from the brightest objects is measured in terms of negative magnitudes. The brightest star, Sirius, is −1.4m; Venus can be as bright as −4m; the Sun is −27m. The faintest stars seen by the unaided eye are about 6m.

This scale is for *apparent magnitude*: it measures how bright things appear to *us*. Denoted by *m*, it depends on two conditions: how bright the source really is, and how far away it is. The intensity of light from a point source falls off inversely as the square of the distance from it. Thus at 100 feet a streetlight will look 16 times brighter than one of the same kind at 400 feet. In the bottom illustration, all three stars could have the same apparent magnitude as seen from Earth, yet differ in *intrinsic* brightness and in distance from us.

Intrinsic brightness is measured by *absolute magnitude*, denoted by *M*, and defined as the magnitude a celestial object would appear to have at a distance of exactly 10 parsecs (about 32 light-years). For any object at distance *r* from us (measured in parsecs), the relation between apparent and absolute magnitudes is:

$$m - M = 5 \log_{10} r - 5$$

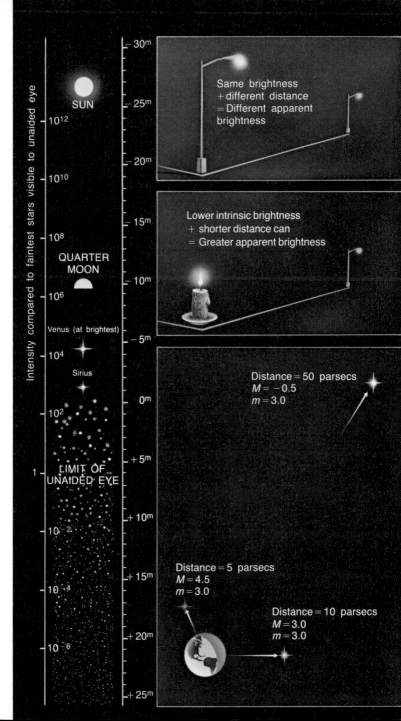

Intensity compared to faintest stars visible to unaided eye

10^{12}	SUN
10^{10}	
10^8	
10^6	QUARTER MOON
10^4	Venus (at brightest)
	Sirius
10^2	
1	LIMIT OF UNAIDED EYE
10^{-2}	
10^{-4}	
10^{-6}	

-30^m
-25^m
-20^m
-15^m
-10^m
-5^m
0^m
$+5^m$
$+10^m$
$+15^m$
$+20^m$
$+25^m$

Same brightness
+ different distance
= Different apparent brightness

Lower intrinsic brightness
+ shorter distance can
= Greater apparent brightness

Distance = 50 parsecs
$M = -0.5$
$m = 3.0$

Distance = 5 parsecs
$M = 4.5$
$m = 3.0$

Distance = 10 parsecs
$M = 3.0$
$m = 3.0$

STAR COLORS AND SPECTRAL TYPE

You may notice that the brighter stars are slightly colored. Every star is of some color, but the fainter the star, the less sensitive the eye is to its color. Color descriptions of stars, moreover, are only approximate.

With the invention of the spectroscope 150 years ago, it became possible to divide the light of any star into its component colors, forming a *stellar spectrum*. Most spectra have a full range of colors, from blue to red, with a few particular colors, or ranges of color, missing. The missing ones, usually designated by the wavelengths of the light, show up as dark lines or bands in the spectrum.

The overall color of a star is an indication of its temperature, just as the dull-red glow of a rod of metal in a fire indicates relatively low temperature; brighter red, a hotter temperature; and orange, yellow, and white, progressively hotter temperatures still. Dark lines result from selective absorption of light by various chemical elements in the star's atmosphere. The lines differ in stars of different temperatures because the different elements absorb light at different temperatures.

Stellar *spectral types*, in order of *decreasing* stellar surface temperature, are designated as O, B, A, F, G, K, and M. Note that only surface temperature is directly observable. The standard mnemonic for spectral types is "Oh, Be A Fine Girl, Kiss Me." Finer divisions are distinguished by dividing each class into ten parts, or sometimes even smaller divisions. Thus a star may have a spectral type of A5 (cooler than A4, hotter than A6) or G2 or B1.5.

Stars are believed to evolve, as described on p. 50. Over the course of time, this evolution affects their spectrum. Stars of types O, B, and A are said to be *early-type* stars, while those of types K and M are said to be *late-type*. A B3 star is said to be "earlier" than a B9. If there are peculiarities in the spectrum, a "p" is placed after the type. (A very few stars are of types denoted by WN, WC, R, N, S. White dwarfs may have a designation such as DA. For more information, consult an astronomy text.)

The *luminosity class* of a star, corresponding to its absolute magnitude, is designated under two schemes:

New Notation	Old Notation	Luminosity Class
Ia, Iab, Ib	c	Supergiants
II		Bright giants
III	g	Giants
IV	sg	Sub-giants
V	d	Dwarf (main sequence)
VI	wd	White dwarf

A complete spectral designation for the star Vega might be given as A0V (read "A-zero-five") or as dA0 (read "dwarf A-zero").

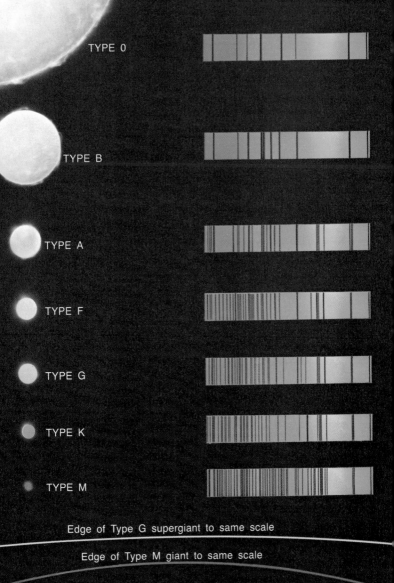

SPECTRAL TYPES
AND SIZES
OF MAIN SEQUENCE
STARS

TYPE O

TYPE B

TYPE A

TYPE F

TYPE G

TYPE K

TYPE M

Edge of Type G supergiant to same scale

Edge of Type M giant to same scale

STELLAR EVOLUTION

Stars form from vast interstellar clouds of gas and dust. As a cloud contracts under its own gravitational pull, it heats up (1) and its atoms move faster. Finally the central part of the cloud detaches from the rest, concentrating to form a *protostar* (2). Parts left behind may form other stars or a planetary system. As the protostar gets denser and hotter, the central region reaches several million degrees, and nuclear fusion occurs—hydrogen atoms combine to form helium, with a release of energy. The now-mature star blows away the remaining gases and settles down to a career on the main sequence (3), lasting most of its lifetime. How long that lifetime will last, how long the star took to reach that stage (in all cases a small fraction of the main-sequence lifetime), and what size, spectral type, and color the star will become depend on the mass of the star. The more massive stars are hotter, larger, and brighter than stars with less mass. They also have shorter lives—a few million years for the hottest O-type stars, about 10 billion years for stars like the Sun, and perhaps tens of trillions for cool, red stars.

Gradually a star runs out of hydrogen fuel at its core, cools, and swells to become a red giant (4). At this stage, stars of spectral type about F5 and cooler simply begin to cool off and shrink, eventually becoming about the size of Earth. Such are the white dwarfs (5), which, consisting largely of atomic nuclei stripped of their electron shells, are incredibly concentrated, typically having masses which on Earth would weigh perhaps 5 tons per cubic inch. Eventually white dwarfs cool to dark cinders.

More massive stars may go through several red giant stages, even evolving into supergiants (6) thousands of times larger than the Sun. During these portions of their lives, they will form chemical elements as heavy as iron in their cores. Most stars become variable in light output in their later stages, and the most massive ones may explode cataclysmically as supernovae (7), leaving behind perhaps a *neutron star* (8) which may be a *pulsar*, or even a *black hole* (9)—a region around a "dead" star so dense and gravitationally attractive that the light cannot escape from it. In the explosion, chemical elements heavier than iron are formed and spewed out into the interstellar material, to be incorporated later in new generations of stars.

Astronomers use the *Hertzsprung-Russell diagram* to describe the life story of a star (and for many other purposes). The diagram is a graph of spectral type (color, or temperature) versus brightness (luminosity). Each point on the graph represents the surface properties of a star, and a line on the graph is the sequence of stages a star goes through as it evolves. The *main sequence* is the location on the graph of those stars that are in the main hydrogen-fusing stage of their lives.

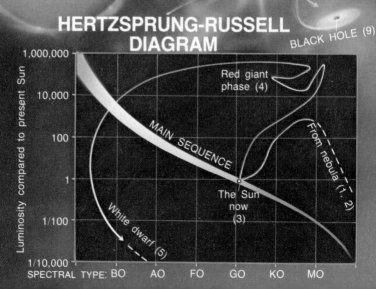

INTERSTELLAR CLOUD (1)

PROTOSOLAR NEBULA (2)

SUN
NOW (3)

SUN
AS WHITE
DWARF (5)

SUN AS RED GIANT (4)

MORE MASSIVE STARS
BECOME SUPERGIANTS (6)

SUPERNOVA (7)
(ONLY VERY MASSIVE STARS)

NEUTRON
STAR
(PULSAR) (8)

BLACK HOLE (9)

HERTZSPRUNG-RUSSELL
DIAGRAM

Luminosity compared to present Sun

1,000,000

10,000

100

1

1/100

1/10,000

MAIN SEQUENCE

Red giant
phase (4)

From nebula (1, 2)

The Sun
now
(3)

White dwarf (5)

SPECTRAL TYPE: B0   A0   F0   G0   K0   M0

MULTIPLE STARS

Our Sun has planets orbiting it. Some stars are orbited by other stars—sometimes several. These are called multiple stars and include binary systems, triple systems, and so on. More than two thirds of all stars have stellar companions.

An *optical double* is a pair of stars that appear close together along our line of sight but are actually very far apart in space. They have no physical connection, gravitational or otherwise.

A *physical double* is a pair of stars gravitationally bound to one another as a system. The same is true of physical triplets, quadruplets, etc. In a system of only two stars, the true orbit of one body about the other lies in a plane, and the shape of the orbit is an ellipse (see Kepler's laws, p. 246). As seen from Earth, the plane of the orbit may be oriented so as to make the orbit appear different from what it really is. The upper illustration shows the true orbit of such a system, and how it would appear from Earth as one star moves about the other. Such a system may even appear edge-on, so that one star periodically eclipses the other (p. 56).

Both members of a physical double revolve about their common center of mass. This obviously is not a visible point. It is thus usual to refer to the less-bright star, called the *secondary*, as orbiting the brighter one, called the *primary*.

Study of a binary, or double star, includes repeated observations, over a span of years, of two measurements: the *angular separation* of the two stars and the direction in the sky of the secondary from the primary. Angular separation is expressed in seconds of arc. Direction, referred to as *position angle* (P.A.), is measured in degrees starting from the north direction (0°) toward the north celestial pole, through east at 90° and around through south, west, and back to north. (P.A. is somewhat analogous to azimuth.) In the eyepiece view shown, the P.A. of the star system is about 250°. Accurately aligned crosshairs in the telescope eyepiece are necessary for such observations. Often another pair of adjustable crosshairs is used to measure the separation. A series of observations of separation and position angle can be plotted to obtain the apparent orbit. Then, from a knowledge of the laws of orbits, the true orbit can be calculated.

A double star's period (the time required for completion of the orbit) may be of almost any length, from days to centuries. Only those stars with relatively wide separations can be seen as separate stars. Very close binaries are not detectable by visual inspection, but their double nature can be deduced from their spectrum. The relative motions of the stars cause shifts in the lines of the spectrum, and from these shifts it may be possible to calculate the orbit. These close double star systems are called *spectroscopic binaries*.

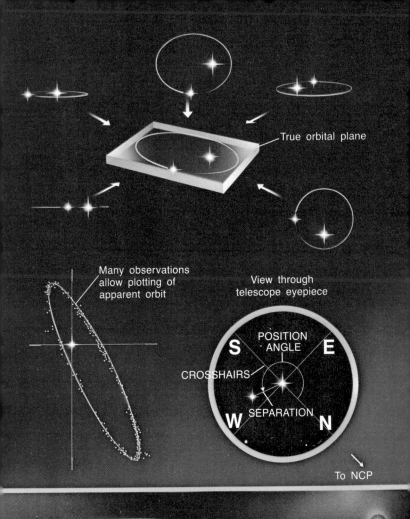

True orbital plane

Many observations
allow plotting of
apparent orbit

View through
telescope eyepiece

S   E

POSITION
ANGLE

CROSSHAIRS

W   N

SEPARATION

To NCP

DOUBLE-DOUBLE
STAR SYSTEM

Multiple star systems make it possible for astronomers to compare the evolutionary processes of stars with a common origin. Eclipsing binaries (p. 56) which are also spectroscopic binaries are extremely valuable in giving astronomers knowledge of the relationship between spectral types and stellar sizes and masses. These fundamental data, obtainable in no other way, are used to validate and refine our theories.

When there are more than two stars in a physical system, the orbits are much more complicated, do not necessarily lie in a single plane, and cannot be represented mathematically by any simple equation. Astronomers wishing to predict the motion in such a cosmic ballet must calculate the motion step by step. Often, however, such a system will consist of a close pair of stars orbited by a more distant star, or even by another close pair. A good approximation of the motions of each pair can be calculated, assuming each is a simple binary, by considering each as a single star.

The nearest star system, Alpha Centauri, is a triple: a G2-type star and a K5 orbit each other closely with a period of 79.9 years, and a third—a red dwarf—orbits the pair. At present this red star is believed to be in a part of its orbit that lies on the side of the pair nearer Earth and thus is the closest star to the solar system; hence its name, Proxima (from Latin *proximus,* "nearest").

A favorite object for small telescopes is Theta Orionis, a quadruple star often called "The Trapezium" because of the arrangement of the components, buried in the haze of Orion's Great Nebula. The four sparkling, very blue stars can be distinguished in binoculars.

In the accompanying list of multiple stars, R.A. is the right ascension; Dec., declination; m_1, apparent magnitude of the primary, and m_2, of the secondary (v if variable); P.A., position angle in degrees; and Sep., separation in seconds of arc. A is the primary; B, the secondary; C, the tertiary.

MULTIPLE STARS FOR BINOCULARS AND SMALL TELESCOPES

Name*		R.A.	Dec.	m_1	m_2	P.A.	Sep.	Comments
ι	Cas	02h24.9m	+67°11′	4.7	7.0	240°	2″.3	A triple. C is 7.1m, 116°, 8″.2
ψ	Dra	17 42.8	+72 11	4.9	6.1	016	2. 3	Yellow and blue
ζ	Cep	22 27.3	+58 10	4v	7.5	192	41. 0	A is prototype of Cepheid variables
55	Psc	00 37.3	+21 10	5.6	8.8	193	6. 6	Orange and blue
γ	Ari	01 50.8	+19 03	4.8	4.8	359	8. 2	Very pretty
γ	And	01 00.8	+42 06	2.3	5.1	063	10. 0	Orange and blue; very pretty
66	Cet	02 10.2	−02 38	5.7	7.7	232	16. 3	Yellow and blue
β	Ori	05 12.1	−08 15	0.2	7.0	206	9. 2	Rigel
λ	Ori	05 32.4	+09 54	3.7	5.6	042	4. 4	Beautiful region
θ	Ori	05 32.8	−05 25					The Trapezium, a close group of 4 stars, with many others, in Orion Nebula
ε	Mon	06 21.1	+04 37	4.5	6.5	027	13. 2	Yellow and blue
β	Mon	06 26.4	−07 00	4.6	4.7	132	7. 4	A is double — 5.2 and 5.6m at 108°, 2″.8. Very nice triple
α	Leo	10 05.7	+12 13	1.3	7.6	307	176. 5	Regulus
γ	Leo	10 17.2	+20°06′	2.6	3.8	122	4. 3	Nice
ζ	UMa	13 21.9	+55 11	2.4	3.9	150	14. 5	Mizar; with Alcor it is a naked-eye double
α	Sco	16 26.3	−26 19	1.2	6.5	274	2. 9	Antares; red and green
α	Her	17 12.4	+14 27	3-4	5.4	109	4. 6	Ras Algethi; red and green
ε¹ Lyr ε² Lyr }		18 42.7	+39 37 {	5.1 5.1	6.0 5.4	002 101	2. 8 2. 3	A double-double; pairs at P.A. 172°, 207″.8
η	Lyr	19 12.1	+39 04	4.5	8.7	082	28. 2	Good at low power
β	Cyg	19 28,7	+27 52	3.2	5.4	055	34. 6	Gorgeous; gold and blue
α¹ Cap α² Cap }		20 14.9	−12 40 {	4.5 3.7	9.0 10.6	221 158	45. 5 7. 1	Optical double B is very close double
α²	Cen	14 33.2	−60 25	−0.1	1.7		17. 6	The nearest star system — a triple. 10m star at 14h26m, −62°28′, called Proxima, is part of system and star closest to Earth.

*For an explanation of star names, see p. 66.

VARIABLE STARS

The light produced by some stars varies in intensity. Repeated observations of the changing magnitude of a variable star over time yield a graph called a *light curve*. Variables are classified according to the differences in their light curves. Some variables vary over regular periods, others are semiregular, while others are completely irregular.

EXTRINSIC VARIABLES: These are stars whose light output is constant, but periodically is partially or completely blocked from our view by another star passing in front. They are also termed eclipsing variables, or Algol-type variables (from Algol, the first such object discovered). Actually, an eclipsing variable is a pair of stars oriented so that Earth lies in or very near the plane of their mutual orbit. Twice during each orbital period there is an eclipse. If the stars differ in spectral type, the eclipse in which the cooler star passes in front of the hotter one is called the *primary eclipse*. When the hotter star passes before the cooler one, a *secondary eclipse* occurs. These points on the light curve are called *primary minimum* and *secondary minimum*, respectively. We cannot observe the two stars separately: their doubleness is deduced from the light variations. A typical period is several days, and the minima usually last minutes to hours.

INTRINSIC VARIABLES: These are stars whose light output changes because of the star's own internal and surface changes. There are many types of such variables, each usually named for the first star of its type identified—the "prototype" star. Opposite, in photo 1, is the variable S Scorpii in a bright phase; in photo 2, taken later, S Scorpii is fainter, but near it is another variable, R Scorpii, now bright enough to see.

Cepheids are among the most numerous and well-known types. These are supergiants that pulsate, changing in surface temperature—hence also in spectral type—over a period ranging from a few days to a few months. A numerical relationship has been observed between the period of pulsation and the average magnitude. Thus, by measuring the period of a Cepheid the astronomer can learn its absolute magnitude. By comparing this with its apparent magnitude he can calculate its distance from Earth. Distances of remote star clusters and galaxies that contain Cepheids can also be measured in this way.

Long-period variables, also called Mira-type variables, are red supergiants with periods of hundreds of days and a great range in magnitude variation. Mira, the prototype, varies in magnitude from 3 to 10 over a period of 322 days.

Novae are stars that explode partially, brightening as they throw off a small percentage of their mass. Then they fade. *Supernovae* are stars that explode cataclysmically, almost totally disintegrating in this final evolutionary stage. Opposite (3) three photos, taken at different times, show brightening and dimming of the Nova of 1910 in Lacerta.

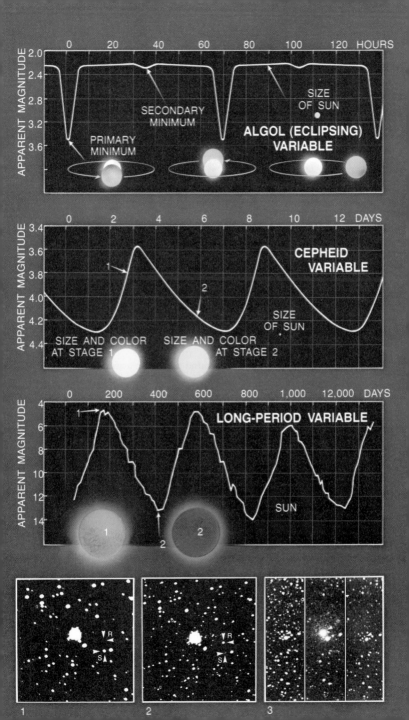

OPEN STAR CLUSTERS

Some stars occur in clusters. Presumably formed from the same cloud of interstellar material at about the same time, they share a common age and composition and are more or less gravitationally bound. The two types are globular clusters (p. 60) and open, or galactic, clusters.

Open clusters are relatively open, irregular groups, consisting of a few dozen to several thousand stars. Even though they appear close together in photographs, the average spacing is several light-years. These are second-generation stars or even younger, formed some time after the original galaxy. Because they were formed in part from debris of older, exploded stars, in which heavy elements had formed, they are relatively rich in heavy elements, in addition to hydrogen and helium. The fact that these stars have similar ages and compositions helps astrophysicists understand how stars' properties change with spectral type, other variable factors being constant.

Still looser groupings of very young stars are called *associations*. Only a few millions of years old, these groups are dispersing. If close to us, they are so spread across the sky that their grouping is not obvious.

At the top of p. 61 is an illustration of what our Milky Way might look like from outside: a flat disk, called the *plane of the galaxy*, a central bulge called the *nucleus*, and the region surrounding all of this called the *halo*. The spiral arms, so hard to detect from our interior vantage point, lie in the plane, and here, too, are found the open clusters. Most of the stars in the plane of the galaxy were formed more recently than those in the nucleus and the halo.

Open clusters include some of the most interesting and beautiful objects in the sky. Several, such as the Pleiades, Hyades, and Double Cluster in Perseus, are visible to the unaided eye. In binoculars or a small telescope they are dazzling. Sometimes they are associated with gaseous nebulae, such as the Great Nebula in Orion.

OPEN CLUSTERS FOR BINOCULARS AND UNAIDED EYES

Name	R.A.	Dec.	Con-stella-tion	Distance, 1,000 lt-yr	Comments
h Persei	02ʰ17.6ᵐ	+ 57°04′	Per	7.0 }	{ Double Cluster in
χ Persei	02 21.0	+ 57 02	Per	8.1 }	{ Perseus
Perseus	03 21	+ 48 32	Per	0.6	
Pleiades	03 45.9	+ 24 04	Tau	0.41	Beautiful!
Hyades	04 19	+ 15 35	Tau	0.13	"Face" of Taurus
Trapezium	05. 34.4	− 05 24	Ori	1.3	In Orion Nebula
Praesepe	08. 39.0	+ 20 04	Cnc	0.59	"Beehive"
Jewel Box	12 52.4	− 60 13	Cru	6.8	Near κ Crucis
M6	17 38.8	− 32 12	Sco	1.5	
M8	18 01.9	− 24 23	Sgr	5.1	In Lagoon Nebula
M11	18 50.0	− 06 18	Sct	5.6	Very rich

The "Double Cluster" in Perseus, h and χ Persei

A sparse open cluster, NGC 7510

A very loose cluster, Abell 5

The Pleiades cluster, showing nebulosity

59

GLOBULAR STAR CLUSTERS

Globular clusters are immense spheroidal, relatively compact group-ings of hundreds of thousands or millions of stars. Formed early in the evolution of the Milky Way galaxy, they surround it in a region called the *galactic halo*, centered over the galactic bulge. Most globulars are very distant from Earth. A few are visible as small fuzzy patches to the unaided eye under good conditions; many are visible in small tele-scopes. The fainter and more distant ones appear only as slightly fuzzy starlike objects.

Globular clusters contain stars that were the first to form when the galaxy was condensing from vast clouds of intergalactic material. This primordial gas contained very little in the way of elements heavier than hydrogen and helium. Most of the gas was used up in making the stars; hence, globulars contain little interstellar material. Because they are isolated from the rest of the galaxy, they were never enriched with more interstellar material, so no second-generation stars formed in globulars as they did in the plane of the galaxy.

In a telescope, globular clusters appear much as they do in photo-graphs, except that a photograph tends to overexpose and blend to-gether the stars near the center. The cluster shown is M13, in the constellation Hercules—a fine cluster for a small telescope.

The table below lists some of the more famous globular clusters.

Name	R.A.	Dec.	Apparent Magnitude, Combined	Con-stella-tion	Dist., 1,000 lt-yr
GLOBULAR CLUSTERS FOR BINOCULARS AND SMALL TELESCOPES					
47 Tucanae	00^{h}23.1^m	$-72°11'$	4.4	Tuc	16
ω Centauri	13 25.6	-47 12	4.5	Cen	17
M3	13 41.3	$+28$ 29	6.9	CVn	35
M13	16 41.0	$+36$ 30	6.4	Her	21
M10	16 56.0	-04 05	7.3	Oph	20
M92	17 16.5	$+43$ 10	6.9	Her	26
M22	18 35.4	-23 56	6.2	Sgr	10
M55	19 38.8	-30 59	6.7	Sgr	20
M2	21 32.4	-00 55	6.9	Aqr	40

SOLAR SYSTEM

MILKY WAY GALAXY

GALACTIC NUCLEUS

GLOBULAR CLUSTER

GLOBULAR CLUSTER

GALACTIC NEBULAE

Gas and dust pervade the disk of our Milky Way galaxy. Concentrations into dense clouds are called *nebulae*. Some are dark, hiding the light of stars behind them; others are luminous, like cosmic neon signs.

Dark nebulae are visible most often as silhouettes against a brighter background. Most of the mottling and seemingly vacant regions in the band of the Milky Way are caused not by a paucity of stars but by absorption of the starlight. Even where the interstellar material is not concentrated into noticeable clouds, it tends to dim and redden the light of more distant stars.

Emission regions, or emission nebulae, are gas clouds—mostly hydrogen like the rest of the universe, mixed with small amounts of helium, oxygen, nitrogen, and other elements. These elements are excited into luminescence by ultraviolet radiation from nearby stars that are hotter than about spectral type B1. The gases absorb the invisible ultraviolet radiation, and reradiate it in the visible part of the spectrum. Such nebulae may contain stars in the process of forming from the gas, or stars just formed. The best-known such nebula is the Orion Nebula, visible to the unaided eye in the "sword" of that constellation.

Reflection nebulae are clouds of interstellar material in which the dust component—of unknown composition—reflects the light from nearby stars. These stars are usually too cool to excite the gases in the nebula to glow, although some nebulae show both emission and reflection properties. The most famous example of a reflection nebula is that surrounding the Pleiades star cluster; the nebulosity is not visible to the unaided eye, but appears when long time-exposure photographs are taken.

Planetary nebulae are so named not because they relate to planets but because they are roughly spherical and some appear greenish, like the outer planets. A planetary nebula is a globe or shell of glowing gases thrown off by a star in a relatively mild explosion late in its evolutionary career. Sometimes the central star that excites the gas to glow can be seen. Because we are looking through a transparent spherical shell of glowing gas, the edges appear brighter than the central regions, and so the nebulae often appear as rings of light. The Ring Nebula in Lyra is the most famous.

Supernova remnants are the chaotic blast debris from very massive stars that have exploded and virtually destroyed themselves. Large optical and radio telescopes have in some cases detected the "corpse" of the star as a neutron star or pulsar at the center of the debris. The Crab Nebula, the result of a supernova recorded in China in A.D. 1054, is the most famous.

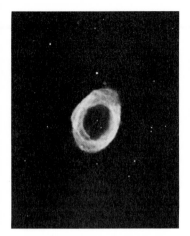

The "Ring" Nebula in Lyra, a planetary nebula. This shell of gas was ejected by the central star

The Great Nebula in Orion, a luminous nebula containing young stars. Note several dark, absorbing regions

SOME GALACTIC NEBULAE

Name	R.A.	Dec.	Constellation	Distance, 1,000 lt-yr	Type/Comments
NGC 1435	03h46.3m	+ 24°01'	Tau	0.4	Reflection nebula; in Pleiades; faint
"Crab"	05 33.3	+ 22 05	Tau	4	Supernova remnant; T
Orion	05 34.3	− 05 35	Ori	1.5	Emission nebula; beautiful; T
"Horsehead"	05.39.8	− 01 57	Ori	1.5	Absorption nebula; near ζ Ori
"Rosette"	06 31.3	+ 04 53	Mon	3	Emission nebula; T
"Owl"	11 13.6	+ 55 08	UMa	12	Planetary nebula
"S"	17 20.7	− 24 59	Oph	?	Absorption and emission nebula; very large; T
"Trifid"	18 01.2	− 23 02	Sgr	3.5	Emission nebula
"Lagoon"	18 02.4	− 24 23	Sgr	4.5	Emission nebula
"Ring"	18 52.9	+ 33 01	Lyr	5	Planetary nebula; T
"Dumbbell"	19.58.6	+ 22 40	Vul	3.5	Planetary nebula

T = Visible in small to medium-size telescope; others visible best by photography

GALAXIES

Galaxies, the largest assemblages in the universe, range from dwarf elliptical galaxies, comparable in size to the largest globular clusters, to the sweeping spirals and the supergiant ellipticals containing thousands of billions of stars.

Spiral galaxies are flat, disk-shaped forms with a central bulge. They are denoted by the type letter *S*, with letters *a*, *b*, and *c* denoting increasing openness of the arms. *S0* galaxies are flat, but show no spiral arms. The Milky Way galaxy (illustration, p. 61; see also pp. 58 and 200) and the Andromeda galaxy are examples of type *Sb*.

Barred spirals also are flat systems, but with spiral arms trailing from a barlike feature which extends through the galaxy's center. They are denoted as type *SB*, again with letters *a*, *b*, and *c* denoting increasing spread between the arms.

Ellipsoidal galaxies range from spherical to football-shaped forms. Whereas spirals contain much interstellar gas and dust, with both young and old stars, ellipsoidal galaxies (sometimes called *ellipticals*) are composed only of old stars, with little or no interstellar material. They are designated type *E*, with numbers running from *0* to *7* to indicate the progression from spherical to spindle-shape. Dwarf ellipticals are the smallest of all galaxies; supergiant ellipticals are the largest.

Irregular galaxies, with no particular shape, designated *I* or *Irr*, contain mostly young stars, with much gas and dust. The Magellanic Clouds, companions of our Milky Way, are examples.

Peculiar galaxies, denoted by the letter *p* after some other designation, are galaxies that depart from the norms. They may show features of several galaxy types, may be strong sources of radio waves, or may have other peculiarities.

The representative galaxies listed below are visible in small and medium-sized telescopes, but don't expect them to look like the long time exposures made with large telescopes. All are far more distant than the stars of the constellations in which we see them.

GALAXIES FOR BINOCULARS AND SMALL TELESCOPES

Name	R.A.	Dec.	Type	Dist., million lt-yr	Con-stella-tion
M31	00^{h}41.6^m	+41°10'	Sb	2.1	And
SMC*	00 52.0	−72 56	Irr	0.2	Tuc
M33	01 32.8	+30 33	Sc	2.4	Tri
LMC*	05 23.7	−69 46	Irr	0.2	Men-Dor
M81	09 53.9	+69 09	Sb	6.5	UMa
M101	14 02.4	+54 26	Sc	14.0	UMa

*SMC = Small Magellanic Cloud; LMC = Large Magellanic Cloud.

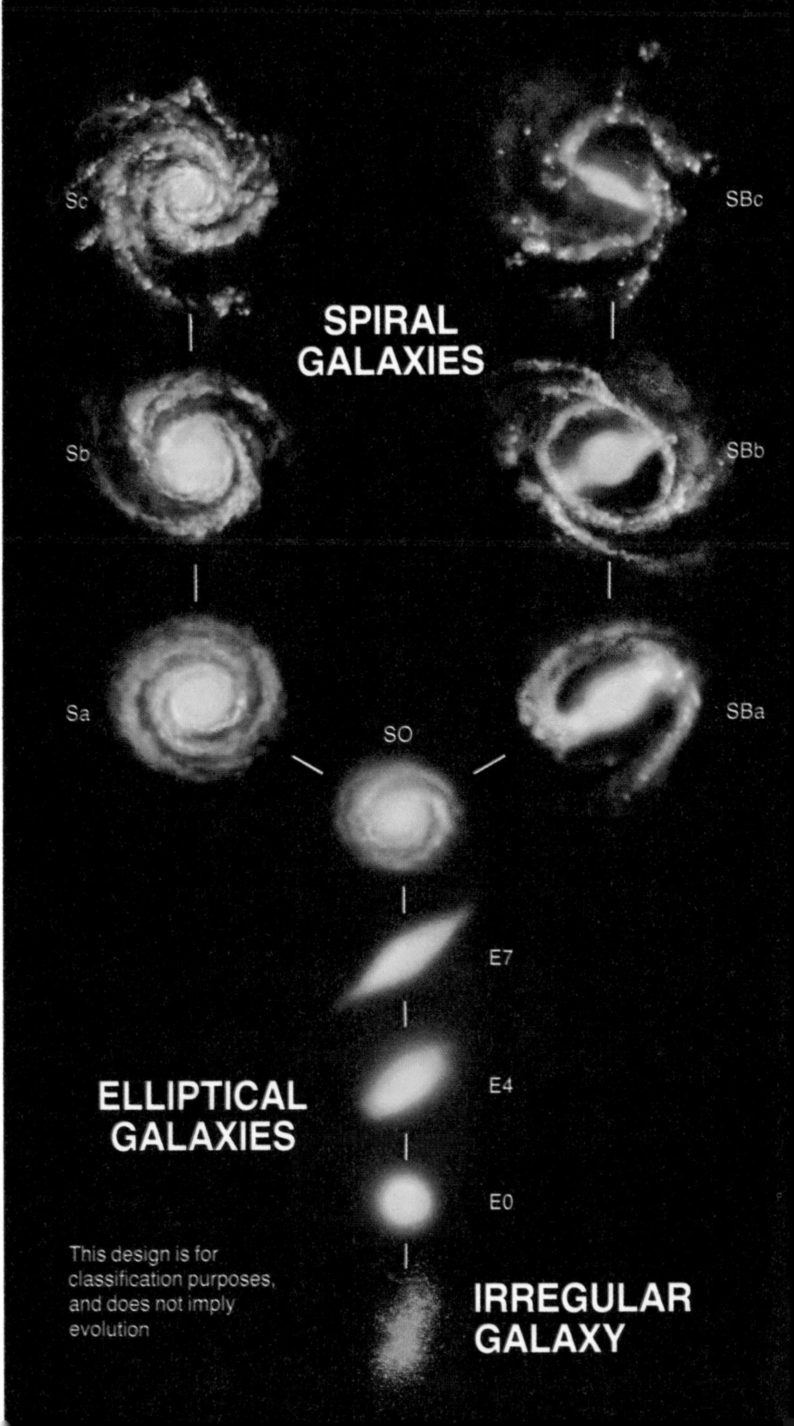

SPIRAL GALAXIES

Sc

SBc

Sb

SBb

Sa

SO

SBa

E7

E4

ELLIPTICAL GALAXIES

E0

This design is for classification purposes, and does not imply evolution

IRREGULAR GALAXY

THE BRIGHTEST STARS

In the following table of the apparently brightest stars, the Bayer designation (from Johann Bayer, 1572-1625, German astronomer) and the common name (if there is one) are followed by coordinates for the year 1900. The apparent magnitude given is what astronomers call "V": a "visual" magnitude based on measurements of the light received by standard equipment. If magnitude is given only to one decimal place, or if a small *v* follows it, the star is variable or otherwise unusual.

The distance to each star is given in light-years (lt-yr). Note that most of the apparently bright stars are both intrinsically bright and rather far away—more than 100 light-years distant. Only a handful are relatively close to us, and most of these are not intrinsically very bright. A table of nearest stars would include very few of the stars listed here.

Absolute magnitude indicates the actual luminosity of the star (p. 46). The spectral type is listed according to classifications previously explained (p. 48). If a star is double (as are two thirds of all stars), with components of nearly equal brightness, each of the pair contributes to the spectrum. Sometimes, too, a single star will show spectral features of more than one type. In both cases, the listing will indicate the composite spectrum—for example, "G + F" for Capella.

In the very brightest stars—those of magnitude brighter than about 1.00—the eye can detect colors, usually rather delicate, when seeing is good. O- and B-type stars will appear bluish white. A-stars will be white, slowly shading to yellow as the spectral type becomes F, then G. Type-K stars will be orange, and M-stars reddish. Among the more noticeable colored stars are Antares, red; Capella, yellow; Sirius, white; and Rigel, blue-white.

Most bright stars have proper names, such as Algol and Betelgeuse. For greater precision, astronomers designate stars by means of Greek letters or Arabic numbers with the name or abbreviation of the constellation in which the star is seen (p. 69). Thus, α Andromedae (or α And) is the star alpha in the constellation Andromeda ("Andromedae" being Latin for "of Andromeda"). Ordinarily, but not always, the Greek letters are used in order of brightness, so that α And is the brightest star in Andromeda, β And the next-brightest, and so on. Other designations refer to their listings in various star catalogs.

THE BRIGHTEST STARS

Bayer Name		Common Name	R.A.	Dec.	Apparent Magni- tude	Dist., lt-yr	Absolute Magni- tude	Spectral Type
α	And	Alpheratz	00ʰ03ᵐ	+28°32′	2.03	127	−0.9	B9
α	Cas	Schedar	00 35	+55 49	2.22	147	−1.0	K0 II
β	Cet	Deneb Kaitos	00 39	−18 32	2.04	59	+0.7	K1 III
β	And	Mirach	01 04	+35 05	2.06	75	+0.1	M0 III
α	UMi	Polaris	01 23	+88 46	2.3v	782	−4.6	F8 Ib
α	Eri	Achernar	01 34	−57 45	0.48	127	−2.2	B5 IV
γ	And	Alamak	01 58	+41 51	2.13	245	−2.2	K3 II
α	Ari	Hamal	02 02	+22 59	2.00	75	+0.2	K2 III
ο	Cet	Mira	02 14	−03 26	2.0v	130	−1.0	M6 III
β	Per	Algol	03 02	+40 34	2.2v	104	−0.3	B8 V
α	Per	Algenib	03 17	+49 30	1.80	522	−4.3	F5 Ib
α	Tau	Aldebaran	04 30	+16 19	0.85	68	−0.7	K5 III
α	Aur	Capella	05 09	+45 54	0.08	46	−0.6	G8 + F
β	Ori	Rigel	05 10	−08 19	0.11	815	−7.0	B8 Ia
γ	Ori	Bellatrix	05 20	+06 16	1.63	303	−3.3	B2 III
β	Tau	El Nath	05 20	+28 32	1.65	179	−2.0	B7 III
δ	Ori	Mintaka	05 27	−00 22	2.19	1,500	−6.1	O9.5 II
ε	Ori	Alnilam	05 31	−01 16	1.70	1,532	−6.7	B0 Ia
ζ	Ori	Alnitak	05 36	−02 00	1.79	1,467	−6.4	O9.5 Ib
κ	Ori	Saiph	05 43	−09 42	2.05	1,826	−6.8	B0.5I
α	Ori	Betel- geuse	05 50	+07 23	0.8v	652	−6v	M2 I
β	Aur	Menkali- nan	05 52	+44 56	1.90	88	−0.2	A2 V
β	CMa	Mirzam	06 18	−17 54	1.98	652	−4.5	B1 II
α	Car	Canopus	06 22	−52 38	−0.73	196	−4.7	F0 Ib
γ	Gem	Alhena	06 32	+16 29	1.93	101	−0.4	AC IV
α	CMa	Sirius	06 41	−16 35	−1.45	9	+1.4	A1 V
ε	CMa	Adhara	06 55	−28 50	1.50	652	−5.0	B2 II
δ	CMa	Wezen	07 04	−26 14	1.84	1,956	−7.3	F8 Ia
α	Gem	Castor	07 28	+32 06	1.58	46	+0.9	M + A
α	CMi	Procyon	07 34	+05 29	0.35	11	+2.7	F5 IV
β	Gem	Pollux	07 39	+28 16	1.15	36	+1.0	K0 III
ζ	Pup	Naos	08 00	−39 43	2.25	2,300	−7	O5
γ	Vel		08 06	−47 03	1.83	489	−4	O7 + DC
ε	Car		08 20	−59 11	1.87	326	−3	K0 + B
δ	Vel		08 42	−54 21	1.95	75	+0.1	A0 V

(Continued on following page)

Note: Coordinates in this table are for the standard year 1900. Coordinates for later dates would be slightly different, because of precession.

Bayer Name		Common Name	R.A.	Dec.	Apparent Magni-tude	Dist., lt-yr	Absolute Magni-tude	Spectral Type
β	Car	Miaplaci-dus	09ʰ12ᵐ	−69°18	1.68	85	−0.4	AC III
ι	Car	Tureis	09 14	−58 51	2.24	650	−4.5	F0 Ib
α	Hya	Alphard	09 23	−08 14	1.99	100	−0.4	K4 III
α	Leo	Regulus	10 03	+12 27	1.35	85	−0.6	B7 V
γ	Leo	Algieba	10 14	+20 21	2.1	108	−0.5	K0 III
α	UMa	Dubhe	10 58	+62 17	1.79	104	−0.7	K0 III
β	Leo	Denebola	11 44	+15 08	2.14	42	+1.58	A3 V
α	Cru	Acrux	12 21	−61 33	0.9v	260	−3.5	B2 IV
γ	Cru	Gacrux	12 26	−56 33	1.64	230	−2.5	M3 II
γ	Cen		12 36	−48 25	2.16	130	−0.5	A0 III
β	Cru		12 42	−59 09	1.26	490	−4.7	B0 III
ε	UMa	Alioth	12 50	+56 30	1.78	82	−0.2	A0 V
ζ	UMa	Mizar	13 20	+55 27	2.09	88	0.0	A2 V
α	Vir	Spica	13 20	−10 38	0.96	260	−3.4	B1 V
η	UMa	Alkaid	13 44	+49 49	1.86	150	−1.6	B3 V
β	Cen	Agena	13 57	−59 53	0.60	114	−5.0	B1 II
θ	Cen		14 01	−35 53	2.06	55	+1.0	K0 IV
α	Boo	Arcturus	14 11	+19 42	−0.06	36	−0.2	K2 III
α	Cen	Rigel Kentaurus	14 33	−60 25	−0.1	4	+4.3	G2 V
β	UMi	Kochab	14 51	+74 34	2.07	104	−0.5	K4 III
α	CrB	Gemma	15 30	+27 03	2.23	75	+0.5	A0 V
α	Sco	Antares	16 23	−26 13	1.0	425	−4.7	M1 Ib
α	TrA		16 38	−68 51	1.93	90	−0.3	K4 III
λ	Sco	Shaula	17 27	−37 02	1.62	325	−3.4	B1 V
θ	Sco		17 30	−42 56	1.87	520	−4.5	F0 Ib
α	Oph	Ras Alhague	17 30	+12 38	2.07	60	+0.8	A5 III
γ	Dra	Eltanin	17 54	+51 30	2.22	117	−0.6	K5 III
ε	Sgr	Kaus Australis	18 18	−34 26	1.83	163	−1.5	B9 IV
α	Lyr	Vega	18 34	+38 41	0.04	26	+0.5	A0 V
σ	Sgr	Nunki	18 49	−26 25	2.08	260	−2.5	B2 V
α	Aql	Altair	19 46	+08 36	0.77	16	+2.3	A7 V
α	Pav	Peacock	20 18	−57 03	1.93	293	−2.9	B3 IV
γ	Cyg	Sadr	20 19	+39 56	2.23	815	−4.7	F8 Ib
α	Cyg	Deneb	20 38	+44 55	1.25	1,600	−7.3	A2 Ia
α	Gru	Al Nair	22 02	−47 27	1.74	68	+0.2	B5 V
β	Gru		22 37	−47 24	2.2v	290	−2.5	M3 II
α	PsA	Fomal-haut	22 52	−30 09	1.16	23	+1.9	A3 V

Note: Coordinates in this table are for the standard year 1900. Coordinates for later dates would be slightly different, because of precession.

THE CONSTELLATIONS

All peoples, from the ancient Sumerians to the medieval Arabs and the American Indians, have had their own sky lore. Celestial objects have been named after gods, people, animals, and tools, and groups of stars have been "connected" to form constellations. The shapes of some constellations suggest the names given to them, such as Draco the Dragon and Orion the Hunter; others, such as Ursa Major, the Great Bear, require a real effort to see any similarity at all. However, the established forms and names of constellations have been convenient, and many that originated in remote antiquity continue in use today.

A number of constellations, as defined on modern astronomical maps, contain some of the same stars that were assigned to them millennia ago. Familiar figures such as Cassiopeia and Orion can be identified on Sumerian tablets, Egyptian pyramids, or Greek statuary thousands of years old. Because stars are so far away and changes in their positions as the eye sees them are likely to become noticeable only over many centuries, the constellations of today have much the same shapes as when they were named by ancient man.

Many star names, such as Betelgeuse and Algol, are Arabic. In the so-called Dark Ages the Arabs, interested in astronomy, gathered up existing astronomical knowledge and added their own.

As the Dark Ages closed, astronomy became an important part of science in the Western world. Observers gave new names to some celestial objects and also named some parts of the sky that had not been previously identified. The new names represented all sorts of things—national leaders, heroes, dogs, optical instruments, and whatnot. Because communication between scientists in those days was poor, there was much confusion as to specific names and as to where one constellation ended and another began. Then, in 1930, the International Astronomical Union defined a set of 88 constellations with precise boundaries. Some of the old names and old constellations—for example, Aeronavigus Pneumatica (The Hot-air Balloon)—were eliminated. One long-recognized constellation, Argo Navis (The Ship), which sprawled over the sky widely, was divided into separate new constellations. Under the new scheme, the whole sky is covered.

Following is a table of constellations now recognized as official by astronomers throughout the world, with the constellation name, the object it is supposed to represent, its area in square degrees (the entire sky encompasses 41,253 square degrees), and the date on which the constellation is on the meridian at 9 p.m.

TABLE OF CONSTELLATIONS

Name*	Common or English Meaning	Size, degrees²	Date on Meridian
Andromeda	Princess	722	Nov 10
Antlia	Air Pump	239	Apr 5
Apus	Bird of Paradise	206	Jun 30
Aquarius	Water Bearer	980	Oct 10
Aquila	Eagle	652	Aug 30
Ara	Altar	237	Jul 20
Aries	Ram	441	Dec 10
Auriga	Charioteer	657	Jan 30
Boötes	Herdsman	907	Jun 15
Caelum	Chisel (Burin)	125	Jan 15
Camelopardalis	Giraffe	757	Feb 1
Cancer	Crab	506	Mar 15
Canes Venatici	Hunting Dogs	465	May 20
Canis Major	Great Dog	380	Feb 15
Canis Minor	Small Dog	182	Mar 1
Capricornus	Sea Goat	414	Sep 20
Carina	Keel (of Argo)	494	Mar 15
Cassiopeia	Queen	598	Nov 20
Centaurus	Centaur	1,060	May 20
Cepheus	King	588	Oct 15
Cetus	Sea Monster (or Whale)	1,231	Nov 30
Chamaeleon	Chameleon	132	Apr 15
Circinus	Compass (or Dividers)	93	Jun 15
Columba	Dove	270	Jan 30
Coma Berenices	Berenice's Hair	386	May 15
Corona Australis	Southern Crown	128	Aug 15
Corona Borealis	Northern Crown	179	Jun 30
Corvus	Crow (or Raven)	184	May 10
Crater	Cup	282	Apr 25
Crux	Southern Cross	68	May 10
Cygnus	Swan	804	Sep 10
Delphinus	Dolphin	189	Sep 15
Dorado	Dorado	179	Jan 20
Draco	Dragon	1,083	Jul 20
Equuleus	Colt	72	Sep 20
Eridanus	River	1,138	Jan 5
Fornax	Furnace	398	Dec 15
Gemini	Twins	514	Feb 20
Grus	Crane	366	Oct 10
Hercules	Strongman	1,225	Jul 25
Horologium	Pendulum Clock	249	Dec 25
Hydra	Water Serpent	1,303	Apr 20
Hydrus	Small Water Snake	243	Dec 10
Indus	(American) Indian	294	Sep 25
Lacerta	Lizard	201	Oct 10
Leo	Lion	947	Apr 10
Leo Minor	Small Lion	232	Apr 10
Lepus	Hare	290	Jan 25

TABLE OF CONSTELLATIONS — Contd.

Name*	Common or English Meaning	Size, degrees²	Date on Meridian
Libra	Scales	538	Jun 20
Lupus	Wolf	334	Jun 20
Lynx	Lynx	545	Mar 5
Lyra	Lyre (or Harp)	286	Aug 15
Mensa	Table Mountain	153	Jan 30
Microscopium	Microscope	210	Sep 20
Monoceros	Unicorn	482	Feb 20
Musca	Fly	138	May 10
Norma	Carpenter's Square	165	Jul 5
Octans	Octant	291	Sep 20
Ophiuchus	Serpent Bearer	948	Jul 25
Orion	Hunter	594	Jan 25
Pavo	Peacock	378	Aug 25
Pegasus	Winged Horse	1,121	Oct 20
Perseus	Hero	615	Dec 25
Phoenix	Phoenix	469	Nov 20
Pictor	Painter's Easel	247	Jan 20
Pisces	Fishes	889	Nov 10
Piscis Austrinus	Southern Fish	245	Oct 10
Puppis	Poop (of Argo)	673	Feb 25
Pyxis	Ship's Compass	221	Mar 15
Reticulum	Net	114	Dec 30
Sagitta	Arrow	80	Aug 30
Sagittarius	Archer	867	Aug 20
Scorpius	Scorpion	497	Jul 20
Sculptor	Sculptor	475	Nov 10
Scutum	Shield	109	Aug 15
Serpens (Caput)	Serpent (Head)	429	Jun 30
Serpens (Cauda)	Serpent (Tail)	208	Aug 5
Sextans	Sextant	314	Apr 5
Taurus	Bull	797	Jan 15
Telescopium	Telescope	252	Aug 25
Triangulum	Triangle	132	Dec 5
Triangulum Australe	Southern Triangle	110	Jul 5
Tucana	Toucan	295	Nov 5
Ursa Major	Great Bear	1,280	Apr 20
Ursa Minor	Little Bear	256	Jun 25
Vela	Sail (of Argo)	500	Mar 25
Virgo	Maiden	1,294	May 25
Volans	Flying Fish	141	Mar 1
Vulpecula	Fox	268	Sep 10

*Pronunciation aids, genitive forms, and abbreviations for constellation names are given with the descriptions of the individual constellations, pp. 112-199.

71

USING STAR CHARTS AND CATALOGS

Charts on the following pages will help you observe with unaided eye, binoculars, or a small telescope. For serious telescope work, more detailed maps are needed (see p. 267). Choose a map that shows stars and other objects not much fainter than objects you will look for. (A map crammed with faint stars can be confusing!) Under good conditions, the eye unaided can detect stars as faint as magnitude 5 or 6; with 7x50 binoculars, magnitude 8 or 9; with a 6-inch telescope, magnitude 13 or 14. Also be sure the scale of the map will be convenient. For work with binoculars, for example, the well-known *Norton's Star Atlas*, with stars to magnitude 6 or 7 and a scale of 3.5 millimeters per degree, will be handier than a sheet from the Mt. Palomar Sky Survey, with its limiting magnitude of 21 and scale of 54 millimeters per degree.

A chart fine for use on a desk may be unwieldy at the telescope. A star atlas, or book of maps, is handier than loose sheets. When consulting charts outdoors at night, preserve visual sensitivity by using minimum light, preferably red.

Charts for beginners usually show stars as constellations, to about 5th magnitude, without coordinates. Any bright star is fairly easily identified by its position relative to other bright stars nearby. Stars fainter than 5th or 6th magnitude are more numerous and thus harder to locate. For them a chart or map with coordinates may be needed. If you have a telescope with setting circles, align it properly relative to the celestial pole and then move the tube so that the coordinates of the sought-for star (call it "Star X") are indicated by the setting circles, with due allowance for the sidereal time (p. 32). Looking into the finder or eyepiece, identify Star X by comparing the field with the chart.

To find Star X with binoculars, first study the chart. Notice the pattern of stars (e.g. triangle, square, loop) in which Star X appears, along with any bright star near it. Notice the pattern's orientation as to the celestial pole. With binoculars locate the pattern, using the bright star (if any) as an aid. In the pattern find Star X.

Use the field of your binoculars for measuring distances across the sky. For example, the field of typical 7x50 binoculars will include a circle of sky about 7° in diameter. If Star X is 14° of arc due north of bright Star Y, it will be twice the width of the circle in that direction. Measure with binoculars accordingly. Note that "due north" means *toward the celestial pole*, and "14° of arc" is equal to 14° *measured along the celestial equator* or along any hour circle.

Relative positions of stars and nebulae change little over decades or centuries. A star chart dated 1950 or even 1900 is sufficiently up to date for ordinary use. For the Moon, planets, and other objects that keep changing their positions against the background of stars, special charts or tables are needed (see pp. 252-263).

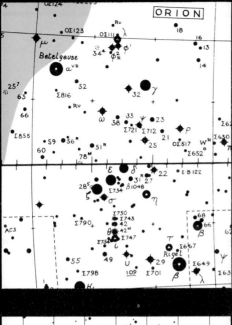

Portions of star charts from two popular star atlases, each reproduced in its original size, showing region around the constellation Orion.

Top: From *Norton's Star Atlas*, which shows 8,400 stars on 16 charts covering entire sky. Limiting magnitude 6^m.5, with occasional fainter stars; scale 3.5mm per degree; shows coordinates, constellations, star names, non-stellar objects, and (in green) the approximate path of the Milky Way.

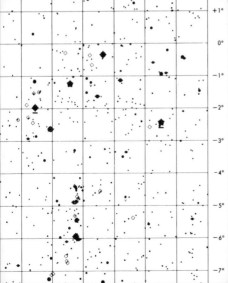

Bottom: From *Smithsonian Astrophysical Observatory Star Atlas*, which shows 258,997 stars on 152 charts covering entire sky. Limiting magnitude 9^m, with occasional fainter stars to 11^m; scale 8.6mm per degree; shows coordinates, but no star names; shows non-stellar objects.

HOW TO USE THE SEASONAL SKY MAPS

On pp. 78-109 are maps of parts of the sky visible from mid-northern latitudes at 10 p.m. March 15, June 15, September 15, December 15. These maps are titled Spring, Summer, Fall, Winter. On each is a list of other times when the sky will appear as shown. For each two weeks earlier that you observe the sky, the time of night is about one hour later. After a full year, the sky as seen at a particular time has gone all the way around.

On pp. 76-77 and 110-111 are maps of the north circumpolar and south circumpolar skies. Each has a table to help in orienting the map properly, and instructions on adjusting for different latitudes.

Keep in mind the effect of changing latitudes or time of observation. North of latitude 40°N., all stars in the NE (northeast) and NW (northwest) quadrants will be a little higher than shown; in the SE (southeast) and SW (southwest), a little lower. South of latitude 40°N., the converse is true.

If you observe at a time later than that on the map, all stars in the NE and SE will be a bit higher and more to the right (south). Earlier, they will be lower and more to the left. Conversely, stars in the SW and NW maps will be lower and more to the right if you observe later, higher and more to the left if you observe earlier.

With the circumpolar charts, when you face north, stars revolve about the celestial pole counterclockwise. When you face south, stars revolve about the south celestial pole clockwise.

On each map the heavy line at the bottom is the horizon. It is curved in this map projection and centered on the directions NE, SE, SW, and NW. The zenith is marked "Z." The map goes beyond the zenith, back over your head, so that you can connect each map with the others of the season. Faint lines parallel to the horizon are lines of constant altitude, every 10°. Lines vertical to the horizon are lines of azimuth. Azimuths are marked along the horizon in degrees, starting at 0° for north and increasing through 90° for east, 180° for south, and 270° for west. The cardinal points (north, east, south, west) and intercardinal points are marked as well.

If you use these charts outside at night, read them with a small flashlight covered with a red filter (e.g. a piece of cellophane). Subdued red light helps to preserve night vision.

When a curved celestial sphere is projected on a flat page, distortions must occur. These are minimal at the center on the horizon, maximal in the upper corners. In the corners the constellation shapes are quite deformed but will help you connect one map with the adjoining one. When facing an azimuth not exactly at one of the intercardinal points, you can rotate the page slightly so that the direction you are facing is down.

Good observing!

HOW TO USE THE CONSTELLATION CHARTS

On pp. 113-199 are charts of 88 constellations, covering the entire sky. They are arranged in alphabetical order, with one exception: Serpens will be found with Ophiuchus rather than under "S," because these two star groups (or three, since Serpens is in two parts) are closely connected both in legend and in the sky. Each page includes two constellations, except that the three small constellations Corona Borealis, Corvus, and Crater are given a single page, and the gigantic Hydra a page of its own.

All charts are to the same scale, with north at top. Each chart shows adjoining constellations and, where helpful, a few nearby bright stars. Coordinates give each constellation's position in right ascension and declination. Each 10° of declination is marked. Each hour of right ascension is given except near the poles, where the lines are close.

On each chart many stars are designated. The most common method of designation is the one devised by Johann Bayer for his atlas of 1603, in which the stars in each constellation are given lower-case Greek letters, usually (but not always) in order of brightness. Some other stars are designated by numbers, called Flamsteed numbers (after John Flamsteed, first Astronomer Royal of England). Regardless of brightness, these numbers increase in order of increasing right ascension within each constellation. Some fainter stars are not named or numbered in our charts, though they do have numbers in some astronomical catalogs. Consult star atlases in the Bibliography.

A few dozen carry proper names. Many are Arabic in origin, for it was the Arabs who preserved and spread classical science around the Mediterranean during the so-called Dark Ages. A few star names are modern. Not all names given are in common use.

Nebulae, galaxies, and star clusters are usually identified by numbers. M numbers are from the catalogue compiled by Charles Messier in 1784-6. NGC numbers are from the *New General Catalogue* prepared by J.L.E. Dreyer, published in 1888.

A list of the 88 constellations appears on pp. 70-71. These are the official groups adopted by the International Astronomical Union in 1930. At that time all the conflicting lists of constellations were worked into the 88 official groups, which include the entire sky and do not overlap. The boundaries were established along lines of constant right ascension and declination, based on the coordinates of 1875. Because of precession, the boundaries move slowly against the background of stars, but this is not a major problem. Most professional astronomers do not rely on constellation designation of stars any more, but use celestial coordinates and star-catalog numbers instead.

Descriptions of the individual constellations accompany the charts. Following the constellation names you will find pronunciation aids, genitive forms, abbreviations, and common English equivalents.

NORTH CIRCUMPOLAR CONSTELLATIONS

These stars are always above our northern horizon, even in daytime when we can't see them. The portion of sky that is circumpolar depends upon your latitude. The north star is as many degrees above the northern horizon as your degrees of latitude. At latitude 50°, all stars within 50° of the pole—that is, down to declination 40°—will be circumpolar.

The chart on the opposite page shows the sky down to declination 40°. At latitude 50°, use the chart as is; at latitude 40°, cover the bottom 10° of the chart; at 30°, cover the bottom 20°; and so on. At the equator, no stars are circumpolar.

The table of dates and times, and the explanation below, will tell you what part of the circumpolar chart should be held down.

Among north circumpolar constellations are some of the best-known star groups, including some recorded in the most ancient times. In our day the north celestial pole is about 1° from Polaris, the so-called north star, at the end of the handle of the Little Dipper, in Ursa Minor. In the time of the ancient Egyptians the bright star closest to the pole was Thuban, in Draco. Five thousand years from now the pole will lie in Cepheus, and 14,000 years in the future the bright star closest to the pole will be Vega, in Lyra.

To find our present north star, look for the large, bright constellation Ursa Major, the Great Bear. In this constellation is the Big Dipper, part of the bear. The two stars in the bowl of the dipper farthest from the handle are Merak and Dubhe, the Pointers, about 5° apart. Draw a line from the one at the bottom of the bowl to the one at the top, and extend the line northward about five times farther (25°) to reach Polaris, the brightest star in its vicinity.

Note: As indicated on p. 72, distances across the sky can be measured in degrees by using the field of your binoculars or telescope eyepiece as the standard. Directions are with reference to the celestial pole.

HOW TO ORIENT THE CHART

To properly orient the chart on the next page: Note that down the left side is a list of dates. Across the top are hours of the night, from 6 p.m. to 6 a.m. Follow the row for the approximate date across to the column with the hour you are observing. The letter of the alphabet at the intersection is the one that should be at the *bottom* of the chart at that hour and date. Because of the motion of Earth around the Sun, the sky shifts by about one hour, or one letter, each two weeks. Use this fact to interpolate between dates. For instance, if observing on January 15 at 11 p.m., hold the chart with "G" at bottom. Remember that these orientations are approximate.

NORTH CIRCUMPOLAR STARS

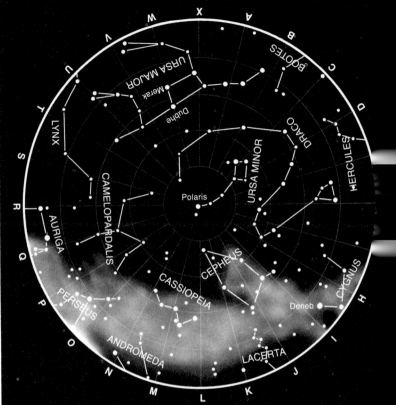

	TIME												
		p.m.							**a.m.**				
	6	**7**	**8**	**9**	**10**	**11**	**12M**	**1**	**2**	**3**	**4**	**5**	**6**
DATE													
Jan 1	A	B	C	D	E	F	G	H	I	J	K	L	M
Feb 1	C	D	E	F	G	H	I	J	K	L	M	N	O
Mar 1	E	F	G	H	I	J	K	L	M	N	O	P	Q
Apr 1	G	H	I	J	K	L	M	N	O	P	Q	R	S
May 1	I	J	K	L	M	N	O	P	Q	R	S	T	U
Jun 1	K	L	M	N	O	P	Q	R	S	T	U	V	W
Jul 1	M	N	O	P	Q	R	S	T	U	V	W	X	A
Aug 1	O	P	Q	R	S	T	U	V	W	X	A	B	C
Sep 1	Q	R	S	T	U	V	W	X	A	B	C	D	E
Oct 1	S	T	U	V	W	X	A	B	C	D	E	F	G
Nov 1	U	V	W	X	A	B	C	D	E	F	G	H	I
Dec 1	W	X	A	B	C	D	E	F	G	H	I	J	K

MAGNITUDES
First ●
Second ●
Third ●
Fourth ·
Fifth ·

SPRING: NORTHEAST

In spring, the central figure in the northeast is Ursa Major, the Great Bear. The Big Dipper, part of the bear, is not a constellation by itself; it is an "asterism." Its most familiar use is in finding the north star, Polaris. Polaris forms the end of the handle of the Little Dipper, in Ursa Minor, the Little Bear. Most of this constellation is faint, not easy to find. Use the Pointers to locate Polaris (p. 76), then follow the chain of fainter stars curving to the Little Dipper's bowl, about 11° away.

Twining between the Bears is Draco. In spring his head is close to the horizon. Just rising nearby, a little to the east, is Hercules. Cepheus, the King, just under Polaris, is faint.

Imagine a long, curved line following the Big Dipper's handle (the Bear's tail) about 25° almost due east. This line is the "arc to Arcturus," the bright yellow-orange star at the end of the arc. It is in the constellation Boötes, the Herdsman (who lies on his side). Between Boötes and Hercules is the beautiful circlet Corona Borealis, the Northern Crown.

Extending the arc of the Bear's tail past Arcturus, you "speed on to Spica," a bright blue-white star just off this map (see p. 81).

Go back to the Great Bear once more—to the Pointers. Extend a line about 40° long approximately southward to the bright blue star Regulus, in Leo the Lion. Leo stands high in the southeast. Now locate some of the faint constellations surrounding Ursa Major and Leo. Almost overhead is Lynx, starting near Leo's head and going northwest. Leo Minor is between Leo and the Big Dipper's bowl.

Underneath the Great Bear's tail (the Big Dipper's handle) are Canes Venatici, the Hunting Dogs. A bit farther south, about a third of the way from the triangle of Leo's hindquarters to Arcturus, lies Coma Berenices, Bernice's Hair—a faint group, mostly stars of 5th and 6th magnitudes.

Lastly, see how much of the Great Bear herself you can find. Her hind legs extend up toward Leo Minor. The front legs and the head are toward Lynx. Ursa Major is very large.

Take binoculars and scan the Milky Way, that bright hazy band stretching from Cepheus northwestward. You are looking from the outer to the inner regions of our galaxy. With unaided eyes, locate Mizar, the star at the bend of the Big Dipper's handle. Try to distinguish little Alcor, very close to it. Mizar and Alcor together constitute a double star. Next, use binoculars to see if you can split Mizar, which is itself a double.

For telescopes, the constellations Virgo, Leo, and Canes Venatici contain several 9th-magnitude galaxies, while the western side of the trapezoid in Hercules contains the famous globular cluster M13. For locations, refer to the individual constellation maps.

Turn to the southeast, locate Leo, and go on to the next map.

GEMINI · Pollux · Castor · AURIGA · LYNX · CANCER · SEXTANS · Regulus · LEO MINOR · LEO · Z · CAMELOPARDALIS · URSA MAJOR · Merak · Dubhe · CANES VENATICI · COMA BERENICES · Polaris · URSA MINOR · Arcturus · BOOTES · CEPHEUS · DRACO · CORONA BOREALIS · HERCULES · SERPENS · EQUATOR · ECLIPTIC

0 N · 15 · 30 · 45 NE · 60 · 75 · 90 E

Local Sidereal Time: 9^h 30^m

Local Mean Time:	Date:
12 midnight	February 15
11 p.m.	March 1
10 p.m.	**March 15**
9 p.m.	April 1

MAGNITUDES
First ●
Second ●
Third ●
Fourth ·
Fifth ·

SPRING NE

SPRING: SOUTHEAST

Center of attraction in the southeast is the anciently recognized constellation Leo, high above the horizon. Leo is the Nemean Lion hunted by Hercules. To find Leo, draw an imaginary line from the Pointers southward to Regulus, the heart of Leo, and to the head, consisting of less-bright stars. The head actually resembles a backward question mark, or sickle. The hindquarters of Leo are a right triangle, and the 2nd-magnitude star at the tail is Denebola, "tail of the lion." Regulus means "the little king."

Almost due east is the bright yellow-orange star Arcturus, at the bottom of kite-like Boötes, the Herdsman. Some people think Boötes looks like an ice-cream cone.

Below Arcturus and to the right is very bright Spica, marking Virgo, the Maiden (Astraea). How much of her can you find? She stretches from near the horizon about halfway to Leo. Later tonight Libra, the Scales of Justice, will rise after Virgo.

The ecliptic, or path of the Sun in the sky, crosses this region. A line drawn from Regulus to Spica is close to the ecliptic. From Spica the ecliptic goes back behind your head through the feet of Gemini and between the face and shoulder of Taurus.

Just to the right of Spica at this time is a group of four stars making up Corvus, the Crow. (Mariners call this group the Sail, because it is shaped like a gaff rig of a sailing ship.) Nearby, just a bit higher, is Crater, the Cup.

Below Crater, Corvus, and Virgo stretches the largest constellation of all—Hydra, the Water Serpent. Most of its stars are faint. Starting at Hydra's head, a circlet of stars in front and below the sickle of Leo, follow a line of stars down and to the left, approximately toward Corvus, but with many twists and turns, as befit a snake. To follow Hydra all the way you may want to refer to its separate constellation map (p. 153).

Between the coils of Hydra and the southern horizon are some obscure constellations. Antlia, the Air Pump, is just below Hydra. Very close to the horizon is the northernmost part of Vela, the Sail—once part of Argo Navis, the Ship, a very large constellation now split up into several smaller groups. Another group, slightly higher, is Pyxis, the Ship's Compass, a little west of due south—looking not at all like a real compass.

High in the south just west of Leo is Cancer, the Crab. The ecliptic goes right through its middle. Cancer has no bright stars, but on a clear night near the center of Cancer you can see Praesepe, the Beehive cluster (see chart, p. 123).

To the right of Cancer, a bit west, is very bright Procyon, in Canis Minor, the Small Dog. And so to the next chart.

90
E

105

120

135
SE

150

165

180
S

Local Sidereal Time = 9ʰ 30ᵐ

Local Mean Time	Date
12 Midnight	February 15
11 P.M.	March 1
10 P.M.	**March 15**
9 P.M.	April 1

MAGNITUDES
First ●
Second ●
Third ●
Fourth ·
Fifth ·

SPRING SE

SPRING: SOUTHWEST

In the spring, at early evening, the stars of winter are just disappearing below the western horizon. This part of the sky is marked prominently by several constellations: Orion, low in the west; Canis Major, in the southwest; and Gemini, high in the west.

The ecliptic runs from Regulus, almost due south of you, through the legs of Gemini, down to the northwestern horizon.

Orion can be used to locate other groups. His belt is now almost parallel to the horizon. A line drawn beyond the belt to the right leads to bright, orange Aldebaran, eye of Taurus the Bull, and the V-shaped Hyades, which make up his face. The Bull lies almost due west.

Orion's belt extended to the left leads to Sirius, the Dog Star, brightest in the sky. When close to the horizon Sirius flashes and twinkles from atmospheric turbulence, and is often reported as a UFO. The rest of Canis Major, the Great Dog, lies to the left of Sirius and below it. Beyond Canis Major, toward the south, is Puppis, the Poop (of the ship Argo Navis).

Almost straight up from Sirius is Procyon, the Little Dog Star, in Canis Minor, the Small Dog. (If you see a dog here, you have imagination!) Between the dogs are the faint stars forming Monoceros, the Unicorn. Through this part of the sky flows the Milky Way, upward to the right, or north.

High in the sky, to the right of Procyon and higher, are Castor and Pollux, the Gemini, or Twins. Castor is slightly higher, to the right. The pair are at this time almost erect relative to the horizon. With some imagination you can identify these two stick-figure "boys," their heads marked by Castor and Pollux, their bodies and feet extending downward toward Orion.

To the right, northwest of Gemini, is imposing Auriga, the Charioteer, an irregular pentagon with one very bright star—Capella, the Goat Star. Auriga is often shown holding a she-goat. Below Capella are "The Kids" (shown on the individual constellation map, p. 119). Note that Auriga shares a star with Taurus, the Bull; the star is one of the Bull's horn tips. Taurus himself is easy to find by the V of his face, bright Aldebaran, the long horns, and the Pleiades cluster. He is said to be charging Orion.

Don't miss a last look at Orion. He'll not be in the evening sky again for many months. Locate the bright, red star Betelgeuse and brilliant, blue Rigel. Pick out the faint stars of Orion's shield, held toward Taurus, and then use binoculars or a small telescope to view the Great Nebula, in Orion's sword just below the belt. This is one of the magnificent sights of the sky.

In and around Canis Major are several open clusters, including M41, just visible to the unaided eye.

Now turn to the northwest, and the next chart.

COMA
BERENICES

URSA MAJOR

LEO MINOR

Z

LYNX

LEO

Regulus

ECLIPTIC

Castor

Pollux

Capella

EQUATOR

CANCER

GEMINI

AURIGA

SEXTANS

Procyon

CANIS
MINOR

Aldebaran

HYDRA

Betelgeuse

ANTLIA

MONOCEROS

ORION

PYXIS

CANIS
MAJOR

Sirius

VELA

PUPPIS

Rigel

LEPUS

ERIDANUS

180
S

195

210

225
SW

240

255

270
W

Local Sidereal Time: 9h 30m

Local Mean Time: Date:
  12 midnight   February 15
  11 p.m.   March 1
  10 p.m.   **March 15**
   9 p.m.   April 1

MAGNITUDES
First   ●
Second   ●
Third   ●
Fourth   ·
Fifth   ·

SPRING SW

SPRING: NORTHWEST

Capella is the point of departure for exploring this part of the sky. It's about halfway up in the northwest. Below Capella, and a little to the left, is Taurus the Bull, sharing a star with Auriga the Charioteer.

On the Bull's shoulder, within about 20° of the horizon at this time, is the beautiful Pleiades cluster, the Seven Sisters. Something like a tiny dipper, this group is quite noticeable for its size, and it has been important in the folklore of groups as diverse as the Aztecs, Druids, and Egyptians. For Tennyson the Pleiades were "like a swarm of fire-flies/tangled in a silver braid." When close to the horizon, they are hard to distinguish individually, and may blend together. Very few people can see all seven of the sisters; most will see six, some five; and a few very sharpsighted observers will pick out as many as fourteen! In binoculars you will see hundreds of stars. Altogether the Pleiades are showpieces of the sky.

To the right, north of Taurus, is Perseus, the hero who saved Andromeda from the Sea Monster. Andromeda herself is not very visible now; only her legs stick above the horizon—a most unbecoming attitude for a princess.

Farther to the north is a group of fairly bright stars in the shape of a "W" on its side. This is Cassiopeia, the Queen—Andromeda's mother. At this time of year she is disappearing from the evening sky, but toward morning will rise again in the northeast. She is not quite circumpolar at latitudes lower than about 45° north.

To Cassiopeia's right, almost due north, is Cepheus the King, who was Andromeda's father. The stars here are not bright, and if the northern horizon is not clear, the lower stars may be hard to find. Cepheus is shaped more like a lopsided shed than a king.

Above Cepheus is Ursa Minor, the Little Bear, with Polaris at the tip of the tail marking the North Pole. Higher still is Ursa Major, the Great Bear, upside down.

For a challenge, look in the region of the sky between Ursa Major, Ursa Minor, Cepheus, and Auriga. With a clear, dark sky you can pick out the faint constellation of Camelopardalis, the Giraffe. If the skies are hazy, this area will look blank.

Above Camelopardalis stretches Lynx. It is said that only the eyes of a lynx can see these faint stars.

In the sky now are two famous and important variable stars: Algol, in the constellation Perseus; and Delta Cephei. For their locations, see the constellation charts—p. 175 and p. 131, respectively.

With pp. 78-85 we have gone once around the spring sky of early evening. Go around once more, trying to recognize the brighter stars and constellations, and to find fainter ones you may have missed. If you are staying outside late, use the charts of the next season to get a headstart.

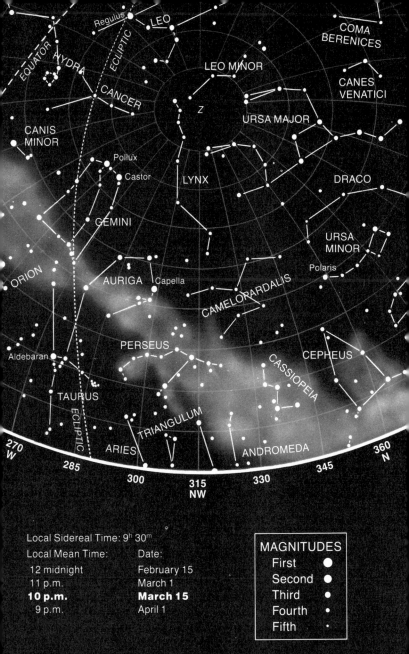

Regulus
LEO
COMA
BERENICES
EQUATOR
HYDRA
ECLIPTIC
LEO MINOR
CANCER
CANES
VENATICI
Z
URSA MAJOR
CANIS
MINOR
Pollux
LYNX
DRACO
Castor
GEMINI
URSA
MINOR
Polaris
ORION
AURIGA
Capella
CAMELOPARDALIS
Aldebaran
PERSEUS
CEPHEUS
CASSIOPEIA
TAURUS
ECLIPTIC
TRIANGULUM
ARIES
ANDROMEDA

270
W
285
300
315
NW
330
345
360
N

Local Sidereal Time: 9ʰ 30ᵐ

Local Mean Time:	Date:
12 midnight	February 15
11 p.m.	March 1
10 p.m.	**March 15**
9 p.m.	April 1

MAGNITUDES
First ●
Second ●
Third ●
Fourth ·
Fifth ·

SPRING NW

SUMMER: NORTHEAST

In the eastern evening sky of summer, three stars are very bright and often seen even through city haze. Almost due east, a quarter of the way up the sky, is Altair, brightest star of Aquila, the Eagle. It resembles an eagle somewhat—or an arrowhead! Slightly higher, and a little toward the north, is Deneb, in Cygnus the Swan. Above Deneb, two thirds of the way from horizon to zenith, is Vega, in Lyra the Harp. Vega, one of the brightest stars in the sky, later at night will pass almost over the heads of those near latitude 38°N. Altair, Deneb, and Vega together are called the Summer Triangle. They will remain in our skies well into the fall, by which time they will be in the western sky in early evening.

While Lyra does not much resemble a harp, or lyre, a small triangle, attached to a small parallelogram, is discernible. Cygnus does resemble its name, Deneb being the tail of the Swan, while a long line of stars pointing south, leading to Albireo, is its beak. A line across the "body" forms the wings. Notice that Cygnus is flying south along the Milky Way. An alternative figure is the Northern Cross, at this time lying on its side, with Deneb at the top of the cross, Albireo at the bottom.

Look more to the north, closer to the horizon, to find Cassiopeia, the Queen, looking like a "W" but slightly lopsided. Above and to the east is Cepheus, the King, her husband.

Due north, and as many degrees of altitude above the horizon as your latitude, is Polaris, the north star, in Ursa Minor, the Little Bear. Just above is Draco, the Dragon. This is a good time to look for him, since he is as high as he will get, and many of his stars are faint. His head is midway up the northeastern sky. His body runs down toward Cepheus, turns northward, loops back up around the Little Bear, and then turns down again.

Just above Draco's head, and above Vega, are the stars of Hercules—a large constellation, as befits the hero. The constellation maps clearly show the "keystone," or "butterfly," in this group.

Now locate some of the more obscure constellations. Between Deneb and Cassiopeia, near Cepheus, is Lacerta the Lizard—not much like its name. Under the Swan's beak, toward Altair, lie the constellations Vulpecula, the Fox, and Sagitta, the Arrow. The Fox is not obvious from its shape, but the Arrow is a fair likeness. Below the Fox and the Arrow, close to the horizon, are two small constellations better seen later tonight, or in later seasons: Delphinus, the Dolphin; and Equuleus, the Colt.

Lastly, use binoculars to scan the region along the Milky Way, from Cassiopeia, Lacerta, and Cepheus, through Cygnus, Aquila, and on to the south. This is one of the most beautiful regions of the sky. As you scan toward the south, turn to the next map.

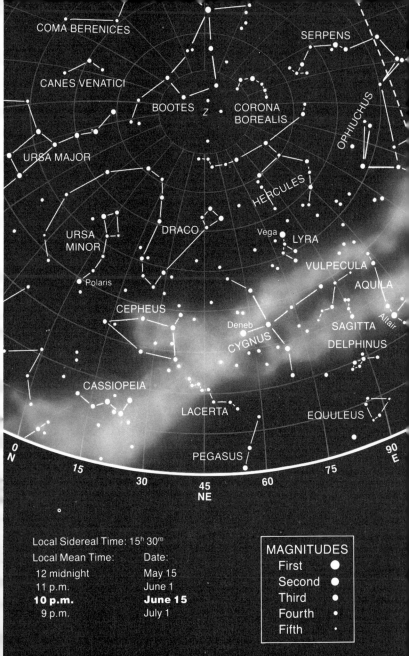

COMA BERENICES
SERPENS
CANES VENATICI
BOOTES   Z   CORONA
BOREALIS
OPHIUCHUS
URSA MAJOR
HERCULES
URSA
MINOR
DRACO
Vega   LYRA
VULPECULA
Polaris
AQUILA
CEPHEUS
Altair
Deneb
SAGITTA
CYGNUS
DELPHINUS
CASSIOPEIA
LACERTA
EQUULEUS
0
N
90
E
15
30
45
NE
60
75
PEGASUS

Local Sidereal Time: 15^h 30^m
Local Mean Time: Date:
 12 midnight   May 15
 11 p.m.   June 1
 10 p.m.   **June 15**
 9 p.m.   July 1

MAGNITUDES
First   ●
Second   ●
Third   ●
Fourth   ·
Fifth   ·

SUMMER NE

SUMMER: SOUTHEAST

To find the main constellations of interest at this time, follow the Milky Way. Look east to the Summer Triangle, where the band of light flows through Cygnus and Aquila. Follow it south to where it is brightest.

Just above the horizon, almost exactly southeast, is Sagittarius, the Archer. He is shooting at the Scorpion to his right. Most people can pick out a teapot among the stars of Sagittarius. Here the star clouds of the Milky Way are brightest, for we are looking toward the center of our galaxy. We cannot see to the center, however, because of obscuring dust that intervenes—dust apparent to us as dark "rifts" in the star clouds.

Whereas more northerly constellations are seen in the sky for several months, southerly constellations such as Sagittarius and Scorpius are seen only briefly, at this time of year. For observers in mid-northern latitudes they are never very high in the sky.

With binoculars or a small telescope, scan the Milky Way and discover what Galileo did in 1610: The Milky Way is the light of millions of stars, most of them too far away and too faint to be seen singly, but blending in a hazy arc. Optical aid reveals that Sagittarius and surrounding regions are rich in star clusters, nebulae, and other interesting sights. For their locations consult the Sagittarius map, remembering that light from some of the globular clusters here has traveled over 20,000 years before reaching your eyes.

The band of the Milky Way continues into Scorpius, one of the most easily recognizable constellations. Red Antares is his heart. Stretching downward toward the horizon are his body and curved stinger-tail; above Antares is his broad head. Scorpius formerly included all the stars just in front of his head that are now part of Libra, the Scales. These stars once formed Scorpius' claws, as indicated by star names in Libra.

The ecliptic passes through Libra, Scorpius, and Sagittarius—all in the zodiac. (Libra, by the way, is the only zodiac figure that is not "alive.") Because of precession, the ecliptic now also passes through a part of Ophiuchus, which lies above Sagittarius and Scorpius. The Sun now spends more time here during each year than in Scorpius.

Ophiuchus, a large constellation, is a man holding a serpent, Serpens. The Serpent is in two parts, at Ophiuchus's sides. The tail, over toward Aquila, is not to be confused with the stars of Scutum, the Shield. The serpent's head is above Libra and stretches upward toward Boötes and Corona Borealis. Ophiuchus is the group in the middle.

Corona Borealis is found easily: a semicircle of stars almost overhead, one of the most attractive small constellations. Nearby is Hercules, above Lyra. More toward the south is Boötes, shaped like a kite, with Arcturus at the bottom.

Turn now to the next map and continue around the sky.

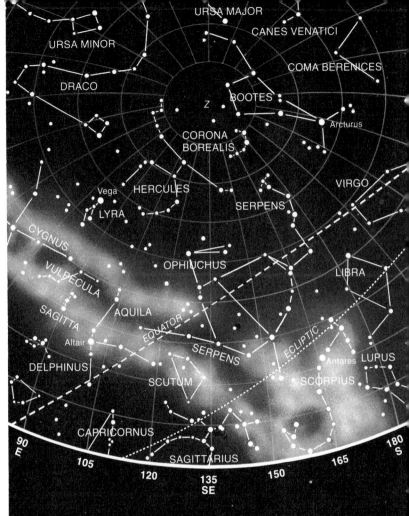

URSA MAJOR
CANES VENATICI
URSA MINOR
COMA BERENICES
DRACO
BOOTES
Z
Arcturus
CORONA
BOREALIS
VIRGO
Vega
HERCULES
SERPENS
LYRA
CYGNUS
OPHIUCHUS
LIBRA
VULPECULA
SAGITTA
AQUILA
EQUATOR
SERPENS
ECLIPTIC
LUPUS
Altair
Antares
DELPHINUS
SCUTUM
SCORPIUS
CAPRICORNUS
90
E
180
S
105
165
SAGITTARIUS
120
150
135
SE

Local Sidereal Time: 15ʰ 30ᵐ

Local Mean Time:	Date:
12 midnight	May 15
11 p.m.	June 1
10 p.m.	**June 15**
9 p.m.	July 1

MAGNITUDES
First
Second
Third
Fourth
Fifth

SUMMER SE

SUMMER: SOUTHWEST

The bright orange star Arcturus, in Boötes, is the starting point for exploring this part of the sky. For many, Boötes is a kite, now stretching right up to the zenith. The kite has a small tail, with Arcturus at the bottom.

Just below Boötes is Virgo, the Maiden, with the bright star Spica. In Virgo is the autumnal equinox, the point where the Sun, on the first day of fall, will cross the celestial equator going south.

Below Virgo are Corvus, the Crow, and Crater, the Cup—constellations better seen earlier in the year. Leo, too, is over in the west, diving to the horizon.

A few stars of the southern constellation Centaurus are just above the southern horizon, visible if skies are clear. Unfortunately the most famous star in this group, Alpha Centauri, the star system nearest to our solar system, never comes above the horizon in mid-northern latitudes. Like the Southern Cross, it is visible only from locations farther south than latitude 25°N. Next to Centaurus is Lupus, the Wolf, filling the region between Centaurus and Scorpius. Above these groups is Libra, which with some imagination resembles a scale, or balance.

Above Libra and Virgo is the head of Serpens the Serpent, pointing toward Corona Borealis, the Northern Crown. It is between Boötes and Hercules, very close to the zenith at this time.

Above Virgo and Leo, below Boötes and Ursa Major, are two constellations: Coma Berenices, Berenice's Hair; and Canes Venatici, the Hunting Dogs. Very clear skies are needed for seeing more than very few stars here. Within Coma Berenices, nearly along the line marked by two stars to the right, is the north pole of our galaxy. This point is 90° away from the plane of the Milky Way, just as our north celestial pole is 90° from the celestial equator.

Within the plane of the galaxy is much obscuring dust. This can prevent us from seeing into great depths of space. However, if we look in the direction of Coma Berenices or Virgo, we are looking out from the thin plane, not through it. Thus a rich field of galaxies is visible. Some of the galaxies are in clusters of galaxies, such as the Coma Cluster or the Virgo Cluster; others are separate. One of the greatest concentrations lies just along the border between Coma Berenices and Virgo. Unfortunately, a telescope of moderate size is required to find most of them. For the more visible galaxies, consult the individual constellation maps, on which "M-objects" are located and briefly described.

Now turn to your right, find the Great Bear, and turn to the next page.

DRACO

URSA MINOR

HERCULES

CORONA
BOREALIS

Z

URSA MAJOR

OPHIUCHUS

BOÖTES

CANES
VENATICI

SERPENS

Arcturus

COMA
BERENICES

ECLIPTIC

VIRGO

LEO

LIBRA

Spica

CORVUS

CRATER

EQUATOR

LUPUS

SEXTANS

HYDRA

180
S

CENTAURUS

195

210

225
SW

240

255

270
W

Local Sidereal Time: 15ʰ 30ᵐ

Local Mean Time:	Date:
12 midnight	May 15
11 p.m.	June 1
10 p.m.	**June 15**
9 p.m.	July 1

MAGNITUDES
First ●
Second ●
Third ●
Fourth ·
Fifth ·

SUMMER SW

SUMMER: NORTHWEST

Ursa Major is the focus in the northwest at this time. She stands nose-down, but in a good location for viewing. The Pointers, Merak and Dubhe, point out Polaris. The handle of the Big Dipper, the tail of the Bear, arcs to Arcturus. The Pointers also direct us to Regulus, in Leo the Lion.

The Bear's front paws, the two stars closest to Lynx, were called by the Arabs Talitha, "third leap of the gazelle," for they were thought to resemble hoofprints. The Bear's hind paws were called Tania, "second leap," and the two stars above Leo Minor were Alula, "first leap."

Test your eyesight by looking at Mizar, next-to-last star in the tail. Good eyes will find two stars here. A telescope will reveal that Mizar, the brighter one, is itself a double. Spectral analysis finds here a third component, too close to the others to be seen separately. Alcor, the Rider, Mizar's principal companion, also is a binary star.

In American Indian legend the stars we call the tail were three hunters chasing the Bear, and Alcor was the pot to cook the Bear in when caught (p. 194). Classical legends said the tail became so long because, when the Bears were causing trouble on Earth, Hercules picked them up by their tails and swung them into the sky.

If you think that is stretching a tale, remember that Draco got hurled into the sky and wrapped around the north pole. Indeed Draco does seem to enclose Ursa Minor, the Little Bear. At this hour the Little Dipper stands upright on its handle.

The 3rd-magnitude star in Draco, about midway between the cup of the Little Dipper and the handle of the Big Dipper, is Thuban, Alpha Draconis, which was the north star about 3,500 years ago. Though fainter than our present north star, Polaris, it was closer to the north celestial pole than Polaris is now or will be at its closest. As explained on p. 26, precession is the reason why different stars become the north star in different epochs.

Leo the Lion, a constellation of spring, is just disappearing in the west. Take one last look at the bright star Regulus, "the Little King," and the famous sickle, or backward question mark. The ecliptic passes very close to Regulus.

Just south of the line marking the lower part of Leo, almost on line from Regulus to Denebola, the tail, are four galaxies. Being of about the 9th magnitude, they require a small telescope, which shows them as hazy blobs of light. Each is a spiral galaxy, somewhat like our Milky Way, but this is apparent only in very large telescopes.

At the time given on the chart, Castor and Pollux are right on the horizon. A little later they will be gone, not to reappear for several months. If you stay out late tonight, use the charts of the next season to find your way around the sky.

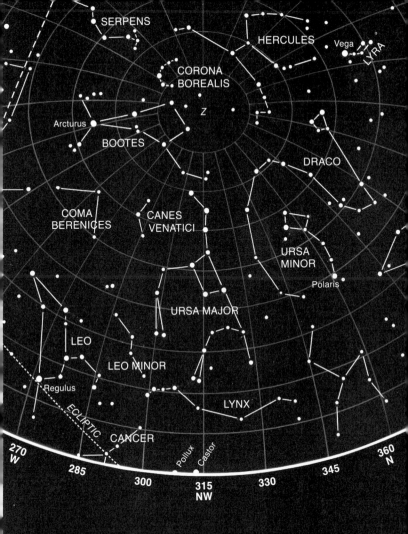

SERPENS

HERCULES

Vega

LYRA

CORONA
BOREALIS

Z

Arcturus

DRACO

BOOTES

COMA
BERENICES

CANES
VENATICI

URSA
MINOR

Polaris

URSA MAJOR

LEO

LEO MINOR

Regulus

LYNX

ECLIPTIC

CANCER

270
W

285

300

Pollux Castor

315
NW

330

345

360
N

Local Sidereal Time: 15ʰ 30ᵐ

Local Mean Time:	Date:
12 midnight	May 15
11 p.m.	June 1
10 p.m.	**June 15**
9 p.m.	July 1

MAGNITUDES
First   ●
Second ●
Third   ●
Fourth ·
Fifth   ·

SUMMER NW

FALL: NORTHEAST

Six of the constellations visible at this time of year are the *dramatis personae* of one of the sky's best-known myths: the story of Perseus and Andromeda (p. 112). Halfway up in the northeast is Cassiopeia, at this time looking like a number 3. She is the cause of the legend, for, renowned for her beauty, she had the audacity to compare herself with the sea-nymphs. Cepheus, Cassiopeia's husband, is above and to the right of Cassiopeia, standing on his head.

Perseus, a third of the way up the northeastern sky, for some people resembles a letter "K" or a fleur-de-lis. He still holds the Gorgon's head in his hand, carefully averting his eyes lest he himself be turned to stone. To distinguish him, look for the three stars in an almost-vertical line. To the right (east) are two lines of stars: the top line is his arm, and the brightest star in the line is Algol, the Demon Star. (The name comes from Al Ghul; hence also our word "ghoul.") Being an eclipsing binary, Algol "winks." About every three days, one star eclipses the other, and for 10 hours the brightness drops from 2nd to 3rd magnitude.

Above Perseus, high in the east, is the Great Square of Pegasus. Two lines of stars running from one corner of the square toward Perseus are Andromeda. Her head is a star shared with Pegasus.

The lines of Andromeda's gown are pairs of stars. Look for the second pair away from the Great Square. Then look slightly north from the fainter one of the pair to another for a faint star. Near this, on a clear night, sharp eyes can see a fuzzy patch of light. This is the Great Galaxy in Andromeda, a spiral 2.1 million light-years from us. It is the most distant object visible to the unaided eye. It is how our Milky Way galaxy would look at the same distance.

This is a very rewarding time of year to scan the sky with 7x50 binoculars. In this part of the sky lie some of the most beautiful and complex patterns of the Milky Way. Starting at the horizon in the northeast, scan upward toward the zenith. Auriga, where the breadth of the Milky Way is small, contains a couple of 6th-magnitude open clusters. In Perseus the hazy band broadens and contains several open clusters, including the famous Double Cluster. Higher in the sky are the rich regions of Cassiopeia, Lacerta, and Cygnus. The direction away from the center of the Milky Way galaxy is toward Auriga, just about on the horizon. The direction of our solar system's revolution around the galactic center is toward Cygnus, overhead. Scanning this quadrant, you are looking toward the outer regions of the Milky Way.

Below Andromeda is the group called Triangulum, the Triangle. Below that is Aries, the Ram, one of the constellations of the zodiac. Above and to the right is Pisces, the Fishes, and below is Cetus, Andromeda's would-be nemesis. These constellations are better seen facing southeast, so—turn the page.

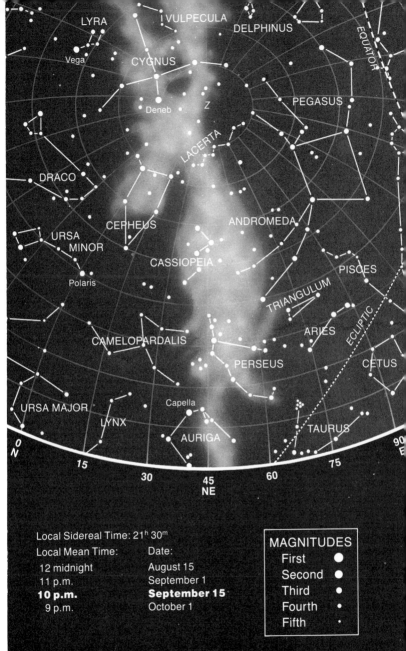

LYRA
VULPECULA
DELPHINUS
Vega
CYGNUS
Deneb
Z
PEGASUS
LACERTA
DRACO
CEPHEUS
ANDROMEDA
URSA
MINOR
CASSIOPEIA
PISCES
Polaris
TRIANGULUM
ARIES
CAMELOPARDALIS
PERSEUS
CETUS
URSA MAJOR
LYNX
Capella
AURIGA
TAURUS
EQUATOR
ECLIPTIC

0
N
15
30
45
NE
60
75
90
E

Local Sidereal Time: 21ʰ 30ᵐ

Local Mean Time:	Date:
12 midnight	August 15
11 p.m.	September 1
10 p.m.	**September 15**
9 p.m.	October 1

MAGNITUDES
First ●
Second ●
Third ●
Fourth ·
Fifth ·

FALL NE

FALL: SOUTHEAST

The southeastern sky at this time is dominated by three large constellations, occupying almost a twelfth of the sky. Closest to the horizon is Cetus, the Sea Monster of the Andromeda legend. It is a gangling group of stars, none very bright. The brightest, Deneb Kaitos, closest to the constellation Sculptor, is only of 2nd magnitude. The head of the monster is the circlet of stars to the east.

Above Cetus is Pisces, the Fishes. The star nearest Cetus is Alrisha, "The Knot," where two strings binding the two fishes are tied. They run up toward Andromeda and westward, under Pegasus. Pisces is best known as the constellation "where spring starts"—that is, where the Sun when moving north crosses the celestial equator. The crossing point is the vernal equinox (p. 24). About 2,000 years ago the vernal equinox was in Aries, but precession has caused the equator to slip westward along the ecliptic. About 600 years hence the vernal equinox will be in Aquarius, farther west. Fall is best for seeing these groups, because then the Sun is opposite them.

Above Pisces is Pegasus, the Winged Horse (his front half, at least). The Great Square is his body, upside down. The lines of stars going up toward Cygnus are his forelegs; the single line ending in a 2nd-magnitude star, Enif, is his neck and nose.

Along the ecliptic west of Pisces is Aquarius, the Water Bearer. Some legends say he was cupbearer to the gods. More often he is pictured as a man with a jar of water on his shoulder, his head marked by the bright star lying almost on the celestial equator. The stream of water flows down, southward, toward the constellation Piscis Austrinus. In this constellation, the Southern Fish, lies the star Fomalhaut. It is of 1st magnitude and the southernmost bright star visible from mid-northern latitudes. (One must be below latitude 30°N. to see Canopus, in the constellation Carina, the next southern bright star.) Because the region of the Southern Fish has few bright stars, Fomalhaut shines conspicuously above the southern horizon.

The huge region from Cetus and Pisces in the east to Pisces Austrinus and Aquarius in the west, and even farther on to Capricornus, was known very long ago as the "watery" part of the sky. Sometimes Eridanus was included. To the Chaldeans this was the Sea, from which the monster-goddess Tiamat arose to battle the powers of light. The progenitors of many of our modern constellations in the region may have been the monsters Tiamat created to aid her.

To the west of Aquarius is Capricornus—and the next page.

Vega
LYRA
CEPHEUS
Deneb
CYGNUS
LACERTA
VULPECULA
SAGITTA
Z
DELPHINUS
Altair
ANDROMEDA
EQUULEUS
PEGASUS
EQUATOR
ECLIPTIC
ARIES
PISCES
CAPRICORNUS
AQUARIUS
PISCIS AUSTRINUS
Fomalhaut
CETUS
SCULPTOR

90
E
105
120
135
SE
150
165
180
S

Local Sidereal Time: 21ʰ 30ᵐ

Local Mean Time:	Date:
12 midnight	August 15
11 p.m.	September 1
10 p.m.	**September 15**
9 p.m.	October 1

MAGNITUDES
First ●
Second ●
Third ●
Fourth ·
Fifth ·

FALL SE

FALL: SOUTHWEST

Now Capricornus is best viewed in early evening. He is supposedly the god Pan, who jumped into the Nile to escape the giant Typhon (p. 126). Capricornus does not resemble a "goat-fish"; it is just a large, triangular shape with slightly curved edges.

Above Capricornus is Aquarius, and above him Equuleus, the Little Horse, a small, faint group of stars just in front of the star Enif, the nose of Pegasus. Next to Equuleus is the constellation Delphinus, the Dolphin. This is the mammal, not the fish, and a little imagination allows us to see a friendly porpoise arching its back out of the water. To many, Delphinus appears as a tiny kite with a tail. Once found, it is easily remembered and located again with pleasure, for there is something charming about this little figure. In mythology it was the dolphin that saved the life of the minstrel Arion when he was forced to jump overboard from a ship in the Mediterranean to save his life (p. 142). Dolphins have long been loved and respected by seamen.

Below and to the right of Capricornus is Sagittarius, the centaur Archer. He will not be above the horizon long at this time of year. Sagittarius is bisected by the ecliptic, the path of the Sun across the background of stars.

Just above Sagittarius is Scutum, a constellation named in modern times. It represents the coat of arms of John Sobieski, a 17th-century king of Poland. It lies below Ophiuchus and Serpens, now partially below the horizon, almost due west.

It is often useful to visualize the celestial equator where it runs through the sky. Its two end points are the due-east and due-west locations on the horizon. Use the chart to trace the equator up through the stars, beginning in Ophiuchus and running through Aquila, between Equuleus and Aquarius, just under the western circlet of Pisces, and then into the southeastern sky through Cetus, to the east. Although the sky "turns around us," the equator will always be in the same place in the sky as seen from any given location. It crosses the meridian—the line that divides the sky into east and west halves—at an altitude that is easily calculated. Subtract your latitude from 90° to get the altitude of the equator where it crosses the meridian. Thus, at latitude 40° in either hemisphere, the equator is 50° above the horizon.

Scanning the Milky Way with binoculars from Cygnus, overhead, to Sagittarius, near the horizon, you are looking toward one of the inner quadrants of our galaxy. The distribution of stars, star clouds, and dark absorption nebulae is very complex.

Fall is the last opportunity to see the bright stars of the previous season: the Summer Triangle. Altair is closest to the horizon; then Vega, halfway up the western sky; and finally Deneb, highest of all. In December only Deneb will be above the horizon in the early evening.

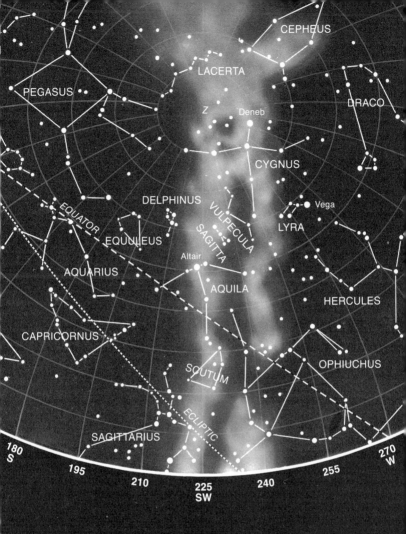

CEPHEUS

LACERTA

PEGASUS

DRACO

Z   Deneb

CYGNUS

DELPHINUS   VULPECULA

EQUATOR   SAGITTA   Vega

EQUULEUS   LYRA

AQUARIUS   Altair

AQUILA

HERCULES

CAPRICORNUS

OPHIUCHUS

SCUTUM

ECLIPTIC

SAGITTARIUS

180
S

195   270
W

210   255

225   240
SW

Local Sidereal Time: 21ʰ 30ᵐ

Local Mean Time:	Date:
12 midnight	August 15
11 p.m.	September 1
10 p.m.	**September 15**
9 p.m.	October 1

MAGNITUDES
First   ●
Second   ●
Third   ●
Fourth   ·
Fifth   ·

FALL SW

FALL: NORTHWEST

The largest constellation, though not the brightest, in this region is Draco. He is the eighth-largest in the entire sky. Now is a good time to see him, for his head lies near Hercules. Unless the horizon is clear, some of his tail stars may be lost in haze.

Lying between the Great and the Little Bear, this dragon is supposed to be Ladon, the one that guarded the golden apples of the Hesperides. Appropriately, the constellation Hercules is nearby, for it was Hercules who slew the dragon to take the apples as one of his Labors. Draco is said also to be the dragon that guarded the sacred spring from which Cadmus had to bring water. When the dragon was finally slain, Cadmus sowed his teeth, and armed men sprang from the ground. Draco's star Thuban was the north star about 2000 B.C.

The Great Bear is "standing" on the horizon. In her is the Big Dipper, all the brighter stars of which, except for the two on the ends, are moving through space in the same direction. They are part of a cluster relatively so close to us that it does not appear as a cluster. Sirius also is a member. The stars in the cluster share a common motion, in the direction approximately along the longest portion of the Dipper's handle, away from the bowl. The two end stars, Dubhe and Alkaid, are moving toward the bottom of the Dipper. In 100,000 years, the bowl of the Dipper will be more like a frying pan, and the handle will be more bent.

Although stars move through space at many kilometers per second, they are so far away that changes in their positions relative to other stars are noticeable only after decades or centuries. Our Sun, with other members of the solar system, is moving too. When looking toward Cygnus, we are looking through the "front windshield" of our solar system, for it is in this direction that we are traveling at 250 kilometers per second as we move around the Milky Way. But we'll never "reach" Cygnus, because it, too, is moving.

Besides this stately revolution around the galaxy's center, each star has a slight "peculiar" velocity relative to its neighbors. Our Sun's peculiar velocity is taking us toward Lyra at the speed of 20 kilometers per second, so "slowly" that it will take us about 15,000 years to cover a mere light-year of space.

Take a pair of binoculars, lie on a cot, and leisurely scan the Milky Way from one horizon to the other. Deneb is nearly overhead. Within Cygnus are bright and dark regions (star clouds and dust clouds) and many nebulae. Southward, in Sagittarius, is the heart of the galaxy; northeastward, toward Perseus, is the route out of our galaxy. The complex distribution of stars, nebulae, and dark rifts that is apparent even to the unaided eye hints at how difficult it is for astronomers to discern clearly the shape and composition of the Milky Way.

EQUULEUS
DELPHINUS
SAGITTA
VULPECULA
LYRA
Vega
CYGNUS
Deneb
Z
LACERTA
CASSIOPEIA
CEPHEUS
OPHIUCHUS
HERCULES
CORONA
BOREALIS
DRACO
URSA
MINOR
Polaris
SERPENS
BOOTES
URSA MAJOR

270
W
285
300
315
NW
330
345
360
N

Local Sidereal Time: 21ʰ 30ᵐ
Local Mean Time: Date:
  12 midnight   August 15
  11 p.m.   September 1
  10 p.m.   **September 15**
   9 p.m.   October 1

MAGNITUDES
First   ●
Second ●
Third   ●
Fourth ·
Fifth   ·

FALL NW

WINTER: NORTHEAST

After being difficult to find for the few months when it is close to the horizon in early evening, Ursa Major, the Great Bear, is now reappearing in the skies shortly after sunset. She is standing on her tail. The Pointer stars, Merak and Dubhe, direct us to Polaris, in Ursa Minor, the Little Bear. In between is the tail of Draco, the Dragon.

Above these groups of stars lies one of the most obscure constellations: Camelopardalis, the Giraffe. Maybe the best way to find it is to look for a region in the northern sky that seems to contain no stars (that is, under city sky conditions). This modern constellation's main feature is that it is long and ungainly like its namesake. Around it are arrayed some of the sky's best-known figures: Cepheus, Cassiopeia, Perseus, Auriga, Ursa Major.

A line of four bright stars runs almost vertically down the eastern sky. Highest is Capella, in Auriga the Herdsman. Somewhat lower are Castor and Pollux, in Gemini the Twins. And just on the horizon is Regulus, in Leo the Lion, becoming higher with each passing minute.

Through Auriga passes part'of the Milky Way, very high in the sky at this time of year and night. Scanning with binoculars or a small, wide-field telescope along this band will reveal the star clouds, clusters, and dark lanes lying toward the outer regions of our galaxy.

The Twins are standing on their heads at this time of year and night. Pollux is closer to the horizon. Even in this awkward position the boys may be found with little trouble. On the ecliptic where it crosses the "feet" of the Twins is the location of the Sun on June 21, when it is farthest from the equator, at the beginning of summer. It was near the stars of the "feet" that Sir William Herschel discovered the planet Uranus, in 1781—the first discovery of a planet in modern times. All the naked-eye planets had been known since antiquity. Strangely, the planet Pluto also was discovered in this same constellation, in 1930. It was near δ Gem (p. 148).

Below Gemini is Cancer the Crab, a famous if indistinct group. Two thousand years ago the Sun was in this constellation at the beginning of summer. Now Cancer is best known as the location of the Beehive Cluster, or Praesepe. This large group of faint stars is located just above the two middle stars in Cancer. In binoculars it is sparkling, in a small telescope even better. A telescope with low magnification (to keep a wide field) is needed for seeing a large part of the cluster all at once.

Leo is now only partly above the horizon, but in an hour all of him will appear, as will more of Hydra. The presence of these constellations in our early evening skies gives notice that although it may be winter, spring will be coming in a few months. By then these stars will be high in the south at sunset.

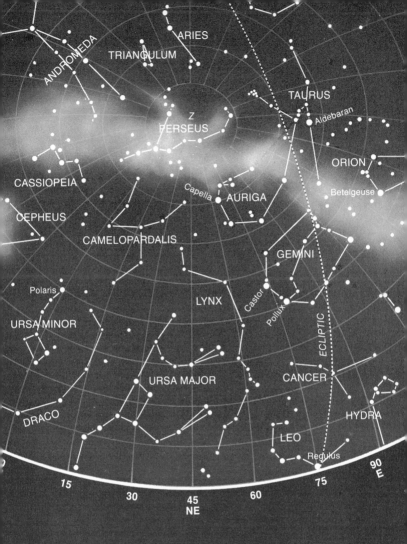

ARIES
TRIANGULUM
ANDROMEDA
TAURUS
Aldebaran
Z
PERSEUS
ORION
Betelgeuse
CASSIOPEIA
Capella
AURIGA
CEPHEUS
CAMELOPARDALIS
GEMINI
Polaris
LYNX
Castor
Pollux
URSA MINOR
ECLIPTIC
URSA MAJOR
CANCER
DRACO
HYDRA
LEO
Regulus
90
E
15
30
45
NE
60
75

Local Sidereal Time: 3ʰ 30ᵐ

Local Mean Time:	Date:
12 midnight	November 15
11 p.m.	December 1
10 p.m.	**December 15**
9 p.m.	January 1

MAGNITUDES
First ●
Second ●
Third ●
Fourth ·
Fifth ·

WINTER NE

WINTER: SOUTHEAST

Orion dominates this part of the winter sky and directs our attention to other bright stars of winter. Most noticeable in Orion is the belt, made up of the stars Mintaka, slightly below the celestial equator; Alnilam, in the middle; and Alnitak, the southernmost. Orion's shoulders are the red Betelgeuse and white Bellatrix; his feet, the blue-white stars Rigel and Saiph. The faint triangle above the shoulders is a beard, and the empty space above that is his head. He holds a curved shield in front of him, toward Taurus, and a club over his head. He is standing on Lepus, the Hare.

Draw a line along Orion's belt toward the southeast. It leads right to Sirius, the Dog Star, brightest star in the sky—brighter than almost any other astronomical object. Of the fixed objects in the sky, only the Sun, Moon, and—at times—Venus and Jupiter are brighter. Canis Major, the Great Dog, is standing on his tail, with Sirius, his eye or nose, topmost. Binoculars will reveal a few open star clusters in this constellation.

East of Sirius is another bright star, almost alone in the sky: Procyon, the Little Dog Star. There are only one or two other naked-eye stars in Canis Minor.

Between the dogs is the faint constellation of the Unicorn, Monoceros. The celestial equator goes right through its star δ Mon.

High above Orion, a little to the east, is large, five-sided Auriga, the Charioteer. The brightest star here is Capella, the She-Goat star. Nearby are "The Kids." Auriga shares a star with Taurus, the Bull.

Taurus was recognized as a constellation thousands of years ago, when the vernal equinox was here. The Bull's face and orange-red eye, the Pleiades, and his two long horns make him quite distinctive. He continually backs away from Orion in a never-ending battle. To find Taurus, use Orion's belt as a pointer toward the west. Aldebaran is the bright orange-red star, part of the V-shaped group called the Hyades. Higher in the sky is the Pleiades star cluster, one of the most lovely star groups of all. Despite its small size the Pleiades cluster, which looks to some observers like a very little dipper, attracts the eye. The viewer with binoculars is greatly rewarded when gazing here.

Near Orion begins one of the longest constellations in the sky: Eridanus, the Celestial River. The star Cursa, not far from Rigel, is the beginning of this meandering pattern, which wends its way toward Cetus, then turns south, circles around Fornax, bends westward again, and finally disappears below our horizon. From there it continues, and to see all of it even under the best conditions you must be south of latitude 30° N.

Eridanus has brought our view around to the south, and to the map on the next page.

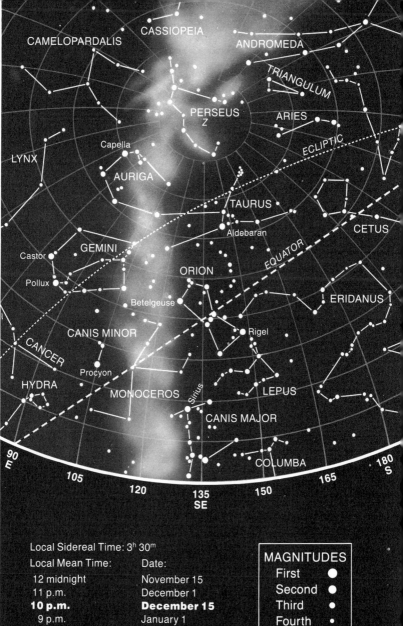

CASSIOPEIA
CAMELOPARDALIS
ANDROMEDA
TRIANGULUM
PERSEUS
Z
ARIES
Capella
ECLIPTIC
LYNX
AURIGA
TAURUS
Aldebaran
CETUS
EQUATOR
GEMINI
Castor
ORION
Pollux
Betelgeuse
ERIDANUS
CANIS MINOR
Rigel
CANCER
Procyon
HYDRA
LEPUS
MONOCEROS
Sirius
CANIS MAJOR
90
E
180
S
105
COLUMBA
120
165
135
150
SE

Local Sidereal Time: 3ʰ 30ᵐ

Local Mean Time:	Date:
12 midnight	November 15
11 p.m.	December 1
10 p.m.	**December 15**
9 p.m.	January 1

MAGNITUDES
First ●
Second ●
Third ●
Fourth ·
Fifth ·

WINTER SE

WINTER: SOUTHWEST

From Eridanus in the south to Pegasus in the west the sky is devoid of bright constellations. Tucked in a loop of the celestial river is Fornax, the Furnace, a modern constellation. Running vertically up the southwestern sky is Cetus, the Sea Monster of Andromeda's story. His head is uppermost, his tail nearer the horizon. Cetus is the fourth-largest constellation in the sky in terms of area. The star midway between the circle of the head and the tail, lying slightly south of the celestial equator, may or may not be visible when you look for it. This is Mira, "The Wonderful," a long-period variable star. At brightest it is of 2nd magnitude, at faintest well below naked-eye visibility, at magnitude 10.

Mira is brightest for a couple of weeks, then over a time span of about eight months slowly fades. Its rise in brightness is much more rapid than its decline. The period is about 331 days, but this too varies, as does the brightness at maximum. Occasionally it will become as bright as Aldebaran. Mira's variability may have been noticed by the ancients, but the first Western astronomer to observe it carefully was David Fabricius, in 1596.

Toward the west lies the Great Square of Pegasus, the Flying Horse. The four corner stars are conspicuous. Trailing upward from the uppermost star of the square is Andromeda, the Princess. Between Pegasus and Cetus are the two circlets of stars forming Pisces, the Fish.

Perseus is directly overhead now. He will be the last of the cast of the Andromeda legend to disappear below the western horizon later tonight. Don't forget to look for Algol, the Demon Star, held in Perseus' right hand. For 10 hours each 3 days the brightness drops from 2nd to 3rd magnitude. When looking at Perseus, you are looking in the direction away from the center of our Milky Way galaxy. Some of Perseus' light is coming from distant stars in our own spiral arm, some from a more distant spiral feature called the Perseus Arm, and perhaps a little bit from the so-called Outer Arm. In this direction our galaxy continues outward for another 15,000 to 20,000 light-years. We are more than 30,000 light-years from the center, for we are in an outer part of the galaxy. Astronomers are not yet certain how fast the number of stars declines with distance from the center, or how sharp the boundary of the galaxy is.

One way of finding out is to look at other, similar galaxies. Two of the closest are both in the sky now. The famous Andromeda spiral is slightly to the north of the second pair of stars in Andromeda away from Pegasus. On a clear night the unaided eye can discern it as a small fuzzy spot. The other close spiral, M33, lies in the constellation Triangulum. Check the constellation map (p. 191) for its exact position; it should be visible in binoculars.

Review this part of the sky, then turn to the next page.

GEMINI

Capella

CAMELOPARDALIS

AURIGA

CEPHEUS

PERSEUS

Z

ORION

Aldebaran

CASSIOPEIA

TAURUS

TRIANGULUM

ARIES

ANDROMEDA

EQUATOR

ECLIPTIC

PISCES

ERIDANUS

PEGASUS

CETUS

FORNAX

δ

AQUARIUS

SCULPTOR

180
S

195

210

225
SW

240

255

270
W

Local Sidereal Time: 3ʰ 30ᵐ

Local Mean Time: Date:

12 midnight   November 15
11 p.m.   December 1
10 p.m.   **December 15**
9 p.m.   January 1

MAGNITUDES
First ●
Second ●
Third ●
Fourth ·
Fifth ·

WINTER SW

WINTER: NORTHWEST

Looking from Perseus, overhead, to Cygnus, near the horizon, we are looking outward through our galaxy in the direction the solar system is traveling as it revolves with other objects around the center of the galaxy. The opposite direction is not visible in midnorthern latitudes because it lies below the horizon in the constellation Vela. The center of our galaxy, toward Sagittarius, is visible at another time of year.

Three distinctive constellation shapes are now visible: the square of Pegasus in the west; the Northern Cross, Cygnus, upright on the northwestern horizon; and Cassiopeia, a distorted letter "W" or "E" halfway up in the northwest.

In November 1572 one of the most important astronomical events in history occurred in Cassiopeia: a "new star" appeared. Its location was a point that makes an almost perfect diamond shape when connected to the three lowest bright stars of the "W" shape. The new object was observed by the famous Danish astronomer Tycho Brahe, who published a detailed report about its changes in brightness. This came at a time when there was much debate over whether the old ideas of the immutability of the heavens were correct. The "stella nova," or "new star," was a nail in the coffin of those old ideas. Forty years later Galileo would seal the coffin when, as the first to use a telescope to study the heavens, he observed satellites revolving about Jupiter—not around Earth, which had been considered the center of the Universe.

The "new star" was actually a very old star blowing up—a supernova (p. 56). Throughout history, only a handful of supernovae have been seen in our galaxy. The more common ordinary novae, much milder explosions, are seen telescopically every year, but only once every decade or so is a nova visible to the naked eye.

Look for δ Cephei, prototype of the invaluable Cepheid variable stars. It ranges in brightness from 4th to 5th magnitude in 5.37 days. It is the star at the top of the tiny triangle on the bottom left of the constellation. To locate it, consult the map for Cepheus.

The north star, Polaris, another Cepheid variable, changes only from 2.1 to 2.2 magnitude, so that you are not likely to notice the change. Its period is about 32 days. The closest bright star to the north celestial pole, about 1° away, Polaris is getting closer, and will be closest in about A.D. 2105. After that, precession will carry it away slowly from the pole.

Extending the curve of the Little Dipper's handle through its bowl, we come to the star Thuban, in Draco the Dragon. Thuban was the pole star about 4,000 years ago.

Now go back and review the entire winter sky. If staying up late, you will be able to use the charts for spring.

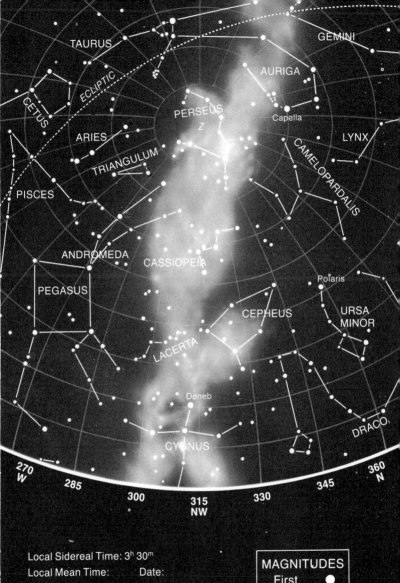

TAURUS GEMINI

AURIGA

ECLIPTIC Capella

CETUS PERSEUS Z LYNX

ARIES

TRIANGULUM CAMELOPARDALIS

PISCES

ANDROMEDA CASSIOPEIA

PEGASUS Polaris

CEPHEUS URSA
MINOR

LACERTA

Deneb DRACO

CYGNUS

270 360
W N
285 345
300 330
315
NW

Local Sidereal Time: 3ʰ 30ᵐ

Local Mean Time:	Date:
12 midnight	November 15
11 p.m.	December 1
10 p.m.	**December 15**
9 p.m.	January 1

MAGNITUDES
First ●
Second ●
Third ●
Fourth ·
Fifth ·

WINTER NW

SOUTH CIRCUMPOLAR CONSTELLATIONS

To a viewer in the northern hemisphere of Earth, there is a region in the northern sky where some constellations never set. These are the stars over the north celestial pole. For the same viewer, there is a region just as large in the southern sky in which some stars never rise. These are the south circumpolar stars. Most cannot be seen from mid-northern latitudes, but some of the more northerly ones are visible in low northern latitudes.

Unlike the north celestial pole, the south celestial pole is unmarked by any bright star. This pole lies in the faint constellation Octans, the Octant. The nearest naked-eye star is of 5th magnitude. One way of finding the pole approximately is to look for the Southern Cross, Crux. If the "upright" of the cross is extended, it will pass almost through the pole. A line perpendicular to α and β Centauri also will come close.

The constellations here are modern, named when European explorers began exploring below the equator. Many different names were given to southern parts of the sky, but few have survived. Some names record new things the explorers found: Dorado, the Dolphin (fish, not mammal); Indus, the Indian; Pavo, the Peacock; Tucana, the Toucan; Volans, the Flying Fish. Some names, such as Octans, the Octant, commemorate modern inventions.

A part of the Milky Way is circumpolar in the southern hemisphere. Many rich star clouds with clusters and nebulae reward patience with binoculars. The two Magellanic Clouds, located in the constellations Tucana and Dorado, look like fragments broken off the Milky Way. Some observers have called them "cirrus clouds," from their superficial resemblance to thin, high clouds in Earth's atmosphere. Originally called the "Cape Clouds," having been seen by explorers rounding the Cape of Good Hope, later they were described by Magellan and named after him. They are in reality small satellite galaxies of our Milky Way galaxy. The Large Cloud (LMC), in Dorado, is 160,000 light-years away. The Small Cloud (SMC), in Tucana and on the border with Hydrus, is 190,000 light-years off. Both appear irregular in shape, but the LMC may be a barred spiral.

HOW TO ORIENT THE CHART

To properly orient the chart on the next page, first note the list of dates down the lefthand side. Across the top are hours of the night, from 6 p.m. to 6 a.m. Follow the row for the approximate date across to the column with the time you are observing. The letter of the alphabet at the intersection is the one which should be at the *bottom* of the chart at that time and date. Because of the motion of Earth around the Sun, the sky shifts by one hour, or one letter, each two weeks. Use this fact to interpolate between dates. For instance, if you are observing on January 15 at 11 p.m., hold the chart with "G" at bottom. These orientations are approximate.

SOUTH CIRCUMPOLAR STARS

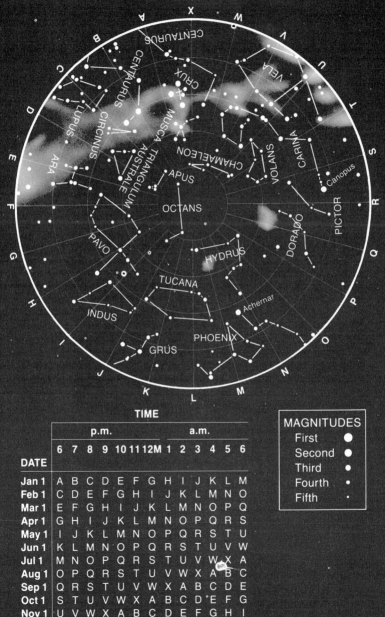

		TIME											
			p.m.						a.m.				
DATE	6	7	8	9	10	11	12M	1	2	3	4	5	6
Jan 1	A	B	C	D	E	F	G	H	I	J	K	L	M
Feb 1	C	D	E	F	G	H	I	J	K	L	M	N	O
Mar 1	E	F	G	H	I	J	K	L	M	N	O	P	Q
Apr 1	G	H	I	J	K	L	M	N	O	P	Q	R	S
May 1	I	J	K	L	M	N	O	P	Q	R	S	T	U
Jun 1	K	L	M	N	O	P	Q	R	S	T	U	V	W
Jul 1	M	N	O	P	Q	R	S	T	U	V	W	X	A
Aug 1	O	P	Q	R	S	T	U	V	W	X	A	B	C
Sep 1	Q	R	S	T	U	V	W	X	A	B	C	D	E
Oct 1	S	T	U	V	W	X	A	B	C	D	E	F	G
Nov 1	U	V	W	X	A	B	C	D	E	F	G	H	I
Dec 1	W	X	A	B	C	D	E	F	G	H	I	J	K

MAGNITUDES

First ●
Second ●
Third ●
Fourth ·
Fifth ·

ANDROMEDA

(ăn-drŏm'-ē-d*a*)   Andromedae   And   *The Princess*

Andromeda was the daughter of King Cepheus and Queen Cassiopeia, rulers of Aethiopia. Because Cassiopeia boasted of her daughter's beauty, the sea god Neptune had Andromeda chained to a sea cliff, there to be sacrificed to Cetus, the Sea Monster; but Perseus used Medusa's head to turn Cetus to stone. The star Alpheratz is Andromeda's head, shared with Pegasus as one of the stars of that constellation's Great Square. Andromeda rises in early evening in the fall.

α is **Alpheratz** (or Sirrah), meaning "the horse's navel," referring to the star's position when it was a part of Pegasus. Known also as "head of the woman in chains," it is of spectral type B9, magnitude 2.03, 127 lt-yr away.

β is **Mirach**, "the girdle" (of Andromeda's gown). This M0III star, magnitude 2.06, is 75 lt-yr away.

γ, **Alamak,** is a triple star close to the galaxy M31 as seen from Earth. The brightest component is of spectral type K3II, magnitude 2.13 at 245 lt-yr. The other two are a very close double star of 5th magnitude, 10″ distant from the K3 II star.

M31 is the spiral galaxy nearest to Earth, the only spiral visible to the unaided eye. A small, faint, fuzzy spot, it was noticed as far back as A.D. 905. It is 2.1 million lt-yr away. From that distance our Milky Way galaxy would look about the same. M31's several hundred thousand million stars shine with light equivalent to that of a 4th-magnitude star. Only a large observatory telescope can resolve M31 into separate stars.

M32, a small ellipsoidal galaxy, is a companion to M31. Another galaxy, **NGC 205,** is nearby. A small telescope reveals these 9th-magnitude galaxies.

ANTLIA

(ănt'-li-*a*)   Antiliae   Ant   *The Air Pump*

This constellation was named by the French astronomer Nicolas Lacaille (1713-62) to honor physicist Robert Boyle's pneumatic machine, used for experiments with gases. Long called Antlia Pneumatica, it became Antlia in 1930 when the International Astronomical Union codified the constellations. Antlia contains no named stars and none brighter than 4th magnitude.

α is an M0 star of magnitude 4.42, 325 lt-yr from Earth.

ε is a star of magnitude 4.64, spectral type M0, at 400 lt-yr.

ι is a G5 star of magnitude 4.70, 230 lt-yr distant.

ANDROMEDA

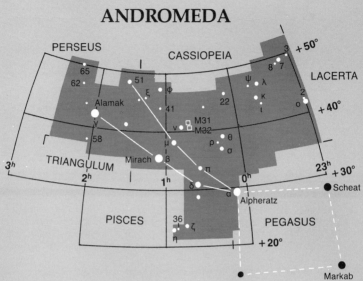

PERSEUS
CASSIOPEIA
LACERTA
+50°
+40°
3ʰ
TRIANGULUM
2ʰ
1ʰ
0ʰ
23ʰ +30°
Alamak
51
ξ
φ
22
ψ λ
κ
ι
o
2
8 7
3
65
62
41
ν M31
M32
θ
ρ
σ
58
μ
Mirach β
π
δ
α Alpheratz
PISCES
36
ζ
η
+20°
PEGASUS
Scheat
Algenib
Markab

ANTLIA

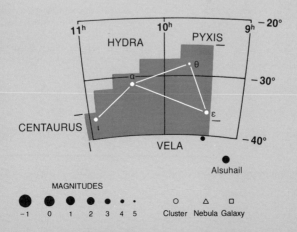

11ʰ
10ʰ
9ʰ
−20°
HYDRA
PYXIS
θ
α
−30°
ι
ε
CENTAURUS
−40°
VELA
Alsuhail

MAGNITUDES
−1 0 1 2 3 4 5   ○ Cluster △ Nebula □ Galaxy

APUS

(ā'-pŭs)   Apodis   Aps   *The Bird of Paradise*

Added as Apus Indica by the German astronomer Johann Bayer (1572-1625), in his atlas of 1603, Apus refers to a long-legged bird of India, not to the more familiar swift family (also *Apus*). Close to the south celestial pole, Apus contains no bright stars.

α is a K5 star of magnitude 3.81. It is 234 lt-yr away.

β, a G8 star of magnitude 4.16, is at a distance of 109 lt-yr.

γ, a subgiant K0 star of magnitude 3.90, is 105 lt-yr off.

AQUARIUS

(*a*-kwâr'-ĭ-ŭs)   Aquarii   Aqr   *The Water Bearer*

Recognized from ancient times, Aquarius is part of the zodiac. The Sun is within its boundaries February 17 to March 13. Aquarius usually is seen as a man pouring water from a jar. His head is the star α; his legs run down to the stars δ and ι. From the jar the stream flows south toward the bright star Fomalhaut, in Piscis Austrinus. The vicinity of Aquarius is the "watery" part of the sky, known to the Babylonians as "the sea." In the so-called "Age of Aquarius," the vernal equinox will be located in this constellation. Contrary to popular song, this era will not arrive for about 600 years (see pp. 26-27). In May, a meteor shower called the "Aquarids" radiates from this region (see p. 228).

α is **Sadalmelik,** meaning "lucky one of the king." A G2Ib star of magnitude 2.93, it is 1,080 lt-yr away.

β, Sadalsud, "luckiest of the lucky," was named for its rising when winter had just passed. At 1,100 lt-yr, it is a supergiant G0Ib star, magnitude 3.07.

γ is **Sadachbia,** "lucky star of hidden things." It is of magnitude 3.97 and spectral type A0. It is 86 lt-yr away.

δ, Skat (perhaps meaning "wish"), is an A3V star 84 lt-yr away, of magnitude 3.28.

ε is **Albali,** "the swallower," an A1 star 172 lt-yr away, of magnitude 3.83.

θ, Ancha, "the hip," is a giant of magnitude 4.32, at 192 lt-yr.

M2 is a globular cluster of 7th magnitude, 40,000 lt-yr away, visible in binoculars.

M72 is a 9th-magnitude globular cluster. Use a small telescope.

NGC 7009, the **Saturn Nebula,** at 3,000 lt-yr, is a planetary nebula. A large telescope shows the 12th-magnitude central star.

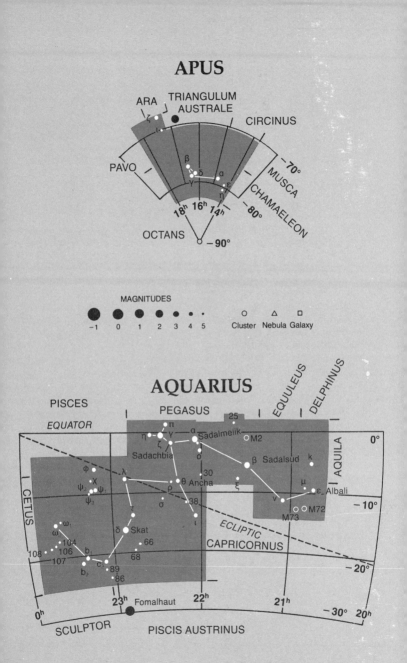

AQUILA

(ăk'-wĭ-la)   Aquilae   Aql   *The Eagle*

This constellation was recognized by several ancient peoples as an "eagle." In ancient Greece it was the eagle that carried the thunderbolts of Jove in his battle with the Titans. Another myth relates that Aquila carried Ganymede, son of the king of Troy, to the sky to serve as cupbearer to the gods, and Ganymede became the constellation Aquarius. Aquila straddles the celestial equator.

α is **Altair,** from the Arabic word for this constellation. This brilliant greenish-white star, of magnitude 0.77, is of spectral type A7V. At a distance of 16 lt-yr, it is the 45th-closest star known.

β, **Alshain,** is 42 lt-yr away. It is of magnitude 3.90 and spectral type dG8.

γ, **Tarazed,** is a 2.72-magnitude star. Of spectral type K3II, it is 340 lt-yr off.

δ is **Denebokab,** "the tail of the eagle." It is 53 lt-yr away and of spectral type F0IV. Its magnitude is 3.38.

ζ is a double (both white, 3" apart) of combined magnitude 2.99 and spectral type A0V. It is 90 lt-yr away.

σ is an eclipsing binary star. The components, B8 and B9, equal 6.8 and 5.4 solar masses and 4.2 and 3.3 solar radii, respectively. The combined magnitude is 5.1. The system lies 137 lt-yr from the solar system.

ARA

(ā'-ra)   Arae   Ara   *The Altar*

This is the altar of the Centaur, near whom it stands. At one time the two constellations Ara and Centaurus were one. At this altar Zeus burned incense to celebrate the victory of the gods over the Titans. None of the stars is named. When seen from the northern hemisphere, Ara is upside down over the southern horizon, with the altar flame "rising" downward.

α, a B2.5V star of magnitude 2.95, is 390 lt-yr away.

β is of magnitude 2.80, spectral type K1.5Ib, and is 930 lt-yr from Earth.

γ is a double star of composite magnitude 3.32, 680 lt-yr away. The main component is a B1I star, 18" from its 12th-magnitude companion.

δ, a B8 star of magnitude 3.79, is 148 lt-yr distant.

AQUILA

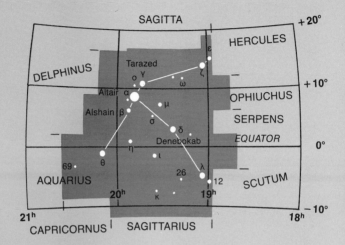

ARA

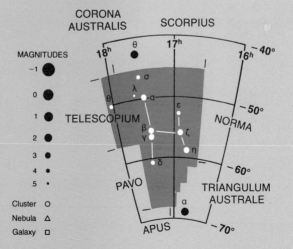

ARIES

(ā'-rī-ēz)   Arietis   Ari   *The Ram*

Sent by Zeus, the Ram rescued Phryxus and Helle, children of the king of Thessaly, from their heartless stepmother. Phryxus later sacrificed the ram, and when he hung its fleece in a sacred grove it turned to gold. Jason and his men, in the ship *Argo*, sought and found the fleece. In ancient times, when the zodiac was defined, the Sun was in Aries on the first day of spring. The vernal equinox is still called the "first point of Aries," though now in Pisces. The Sun is in Aries April 19 to May 15.

α is **Hamal**, Arabic for "sheep," a K2III star, 75 lt-yr away, with a magnitude of 2.00.

β is **Sheratan**, "a sign" in Arabic. It marked the vernal equinox at the time of Hipparchus, who discovered the precession of the equinoxes. It is 52 lt-yr away, 2.65 magnitude, spectral type A5V.

γ, **Mesarthim** (Hebrew for "minister") is a double star for a small telescope. The components, 8" apart, are A0 stars of 5th combined magnitude, 148 lt-yr away.

δ, **Botein**, is "the belly," from an early depiction of the Ram. Now on the tail, it is a gK2 star, magnitude 4.53, distant 172 lt-yr.

AURIGA

(ô-rī'-ga)   Aurigae   Aur   *The Charioteer*

Recognized since early times, displaying many fine objects, Auriga may represent Erichthonius, fourth of the early kings of Athens, whose lameness inspired him to invent the chariot. Or it may represent Neptune in his chariot, rising from the sea. Auriga is often shown as a herdsman holding a goat, with two kids nearby. Its star β is shared with Taurus as one of the Bull's horns.

α, **Capella**, the She-Goat star, honors the goat that suckled the infant Zeus. This is a spectroscopic binary, the primary being a G8III star, slightly variable at about 0.08 magnitude, 46 lt-yr away. The companion is of type F. The pair make up the 6th-brightest star in the sky.

β, **Menkalinan**, "the shoulder of the rein-holder" (charioteer), is of magnitude 1.90, spectral type A2V, at 88 lt-yr.

ε, one of the "Kids," is extremely luminous. Of spectral type F0Iap, at 3,400 lt-yr, it shines with a magnitude of 3.0. It is slightly variable.

η and ζ are the other "Kids."

M36 and **M38** are 6th-magnitude open clusters. Use binoculars.

M37 is an open cluster of 6th magnitude, 4,200 lt-yr away—fine in a small telescope.

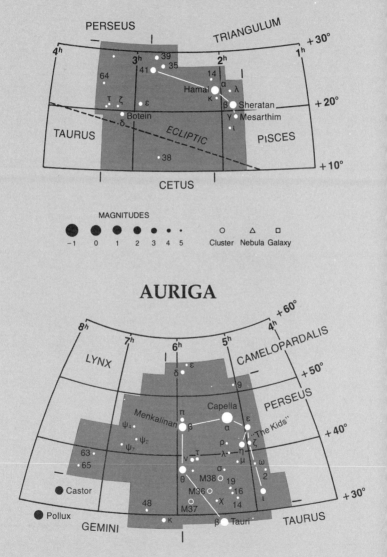

ARIES

PERSEUS | TRIANGULUM

+30°

4ʰ   3ʰ   2ʰ   1ʰ

39
35
41   64
14
Hamal   α   λ
τ   ζ   κ   β   Sheratan
ε   γ   Mesarthim
Botein   δ   ι
TAURUS   ECLIPTIC   PISCES

+20°

38

+10°

CETUS

MAGNITUDES

● ● ● ● ● • ·   ○   △   ▢
-1 0 1 2 3 4 5   Cluster Nebula Galaxy

AURIGA

8ʰ   7ʰ   6ʰ   5ʰ   4ʰ   +60°

LYNX   CAMELOPARDALIS

δ   ε   +50°
9
PERSEUS
Capella
Menkalinan   ε   "The Kids"
π   +40°
ψ₁   β   α   ζ
ψ₃   ρ
ψ₇   λ   η   ω
63   v   τ   μ   2
65   σ   ι
Castor   θ   M38   19
M36   16
Pollux   48   M37   χ   14   +30°
GEMINI   κ   β Tauri   TAURUS

BOÖTES

(bō-ō'-tēz)   Boötis   Boo   *The Herdsman*

Boötes, son of Jupiter and Callisto, invented the plow. The constellation, referred to in Homer's *Odyssey*, was often called the "Bear Driver," being near Ursa Major and Ursa Minor. To the early Greeks it was Lycaon, a wolf; to the Hebrews, a barking dog. Usually Boötes is depicted as a man, but to modern eyes may be a kite or an ice-cream cone, with Arcturus at the bottom.

α is **Arcturus,** 4th-brightest star in the sky. It has been called the Watcher of the Bear, being near Ursa Major. With a magnitude of -0.06, it is 36 lt-yr away. A K2 III star, it is yellow-orange to the unaided eye.

β, **Nekkar,** is the head of Boötes. Of magnitude 3.48, spectral type G8 III, it is 140 lt-yr away. The name is Arabic for the entire constellation.

γ, **Seginus,** is of magnitude 3.05, type A7 III, at 118 lt-yr.

δ is a supergiant G8 III, 140 lt-yr away, magnitude 3.47, with an 8th-magnitude companion 105" distant.

ε, **Izar,** the "girdle," is a triple star. Two members form an orange and green double 3" apart, of types G8 V and K5, respectively 85% and 75% as massive as our Sun, revolving about a common center in 150 years. The system is at 103 lt-yr.

η, **Muphrid,** the "solitary star," is a 2.69-magnitude G0 IV star, 32 lt-yr away.

μ, **Alkalurops,** is "the herdsman's staff." It is of magnitude 4.47 and spectral type A7, at a distance of 988 lt-yr.

CAELUM

(sē'-lŭm)   Caeli   Cae   *The Chisel,* or *Burin*

This constellation was defined by Lacaille, the 18th-century French astronomer. Between Columba and Eridanus, in mid-northern latitudes, it is just visible in winter above the southern horizon. The Chisel is the tool of Sculptor, a constellation west of Eridanus.

α is an F1 star of magnitude 4.52, at 72 lt-yr.

β, an F0 star of magnitude 5.08, is 69 lt-yr. away.

γ is a K5 star, magnitude 4.62, at 230 lt-yr.

δ is a B3 star of magnitude 5.16, distance undetermined.

BOOTES

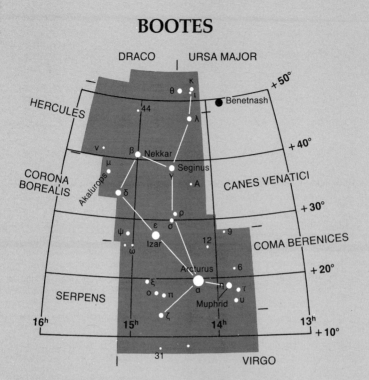

DRACO   URSA MAJOR

HERCULES

κ
θ
ι
+50°
Benetnash
44
λ
+40°
ν
β Nekkar
CORONA
BOREALIS
μ
γ Seginus
CANES VENATICI
Akalurops
δ
A
+30°
ρ
ψ
ε
σ
12
9
COMA BERENICES
Izar
ω
6
+20°
Arcturus
η
τ
ξ
α
u
SERPENS
o
π
Muphrid
16ʰ
ζ
15ʰ
14ʰ
13ʰ
+10°
31
VIRGO

CAELUM

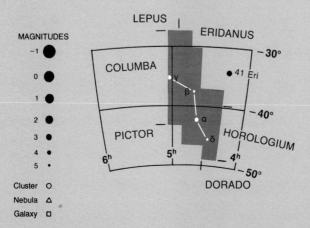

LEPUS   ERIDANUS

MAGNITUDES

−1 ⬤
0   ⬤
1   ●
2   ●
3   •
4   ·
5   ·

Cluster ○
Nebula   △
Galaxy   □

COLUMBA
−30°
γ
41 Eri
β
PICTOR
α
−40°
δ
HOROLOGIUM
6ʰ
5ʰ
4ʰ
−50°
DORADO

CAMELOPARDALIS

(ka-mĕl'-ō-pär'-da-lĭs)   Camelopardalis   Cam   *The Giraffe*

Defined by Jacob Bartsch in 1614, the giraffe has also been called the "camel-leopard." The spelling "Camelopardus" is an old form. There are no very bright stars in Camelopardalis. The famous Perseid meteors radiate from where Camelopardalis adjoins Perseus.

α is a star of magnitude 4.38, type 09, at 3,400 lt-yr.

β is a supergiant of spectral type G2, magnitude 4.22, at 1,700 lt-yr.

CANCER

(kăn-śēr)   Cancri   Cnc   *The Crab*

Famous but inconspicuous, this Zodiac constellation lies just in front of Leo's head. Mythology relates that while Hercules was fighting Hydra, the goddess Juno, who hated Hercules, sent a crab to bite and distract him. Hercules crushed it, but Juno placed it in the sky for loyal service. Some ancients called this star group the Gate of Men, through which souls descended from heaven into newborn babies. Sometimes the Crab is incorrectly depicted as a lobster. Today the Sun is in this constellation July 7 to August 11, but 2,000 years ago the summer solstice occurred here. The tropic of Cancer is the northernmost limit of the Sun's travel, 23½ degrees from the equator.

α is **Acubens,** Arabic for "claws." It is of magnitude 4.27, spectral type dF0, 99 lt-yr away.

β is sometimes called **Al Tarf.** It is of magnitude 3.76, spectral type gK4, and 217 lt-yr away.

γ and δ are **Asellus Borealis** and **Asellus Australis,** the Northern and Southern Asses, ridden by Bacchus and Silenus in their battle with the Titans. γ is of 4.73 magnitude, an A0 star, 233 lt-yr away, while δ is a 4.17-magnitude gK0 star, at 217 lt-yr.

M44 to the unaided eye is a small, faint, fuzzy patch. In binoculars it is a beautiful open cluster. Its name is Praesepe, or "Beehive." Rarely, it is called "the manger" (referring to the Northern and Southern Asses). This cluster, 515 lt-yr off, occupies 1½° of sky. It contains about 100 stars some 400 million years old.

M67, for a small telescope, is an open cluster of about 80 stars, at 2,700 lt-yr, formed perhaps 4,000 million years ago.

122

CAMELOPARDALIS

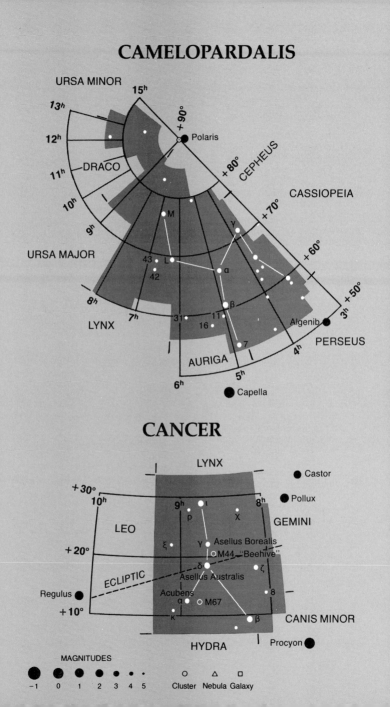

CANCER

MAGNITUDES

-1 0 1 2 3 4 5

Cluster Nebula Galaxy

CANES VENATICI

Defined by the 17th-century Polish astronomer Johannes Hevelius, The Dogs, or Hounds, are Asterion ("starry") and Chara ("beloved")— greyhounds held in leash by Boötes, pursuing the Bear.

α is **Cor Caroli,** "heart of Charles," named by Halley for Charles II of England. It is a double, the primary being a B9.5 star of about 2.90 magnitude, 118 lt-yr away. The 6th-magnitude companion is 20" distant from it.

β is **Chara,** sometimes called "Asterion," a dG0 star of magnitude 4.32, 30 lt-yr away.

M3 is a 7th-magnitude cluster 35,000 lt-yr distant.

M51, the famous **Whirlpool Galaxy,** is only a splotch of light, without detail, in small telescopes. It is of 8th magnitude, type Sc, 14 million lt-yr off.

M63 is an Sb galaxy, 9th magnitude, at 14 million lt-yr.

M94 is a 9th-magnitude Sb-type galaxy at 14 million lt-yr.

CANIS MAJOR

The Great Dog and the Small Dog, east of Orion, join his hunts. Sirius, the Dog Star, brightest in the sky, was carefully observed by the ancient Egyptians, because about the date when it rose at dawn the Nile would flood the countryside. "Dog Days" are a hot period in summer when Sirius, high in the daytime sky, supposedly added to the Sun's heat.

α, Sirius, the **Dog Star,** called by the Egyptians Sothis, is the brightest object in the sky except the Sun, Moon, Venus, and Jupiter. It is of magnitude -1.45, spectral type A1V. At 9 lt-yr it is the 5th-closest star to Earth. Its 9th-magnitude companion, the Pup, 10".3 away, is a white dwarf orbiting Sirius every 49.9 years.

β is **Mirzam,** "the announcer" of the rising of Sirius. Of type B1II, 652 lt-yr away, it is a prototype of the β Canis Majoris variable stars. Its magnitude is slightly variable near 1.98.

γ, Muliphen, is a magnitude 4.07 star of spectral type B8, at 325 lt-yr.

δ is **Wezen,** "weight." Extremely luminous, this F8Ia star, magnitude 1.84, is 1,956 lt-yr away.

ε is **Adhara,** "the virgins," of type B2II, 1.50 magnitude, at 652 lt-yr. An 8th-magnitude companion is 8" away.

ζ, Furud, "bright single one," is of magnitude 3.04, spectral type B2.5V, 390 lt-yr distant.

M41, a 6th-magnitude open cluster visible to the naked eye, is 2,100 lt-yr away.

124

CANES VENATICI

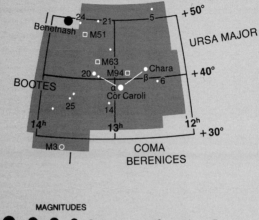

CANIS MAJOR

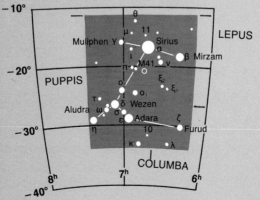

CANIS MINOR

(kā'-nǐs mī'-nēr)   Canis Minoris   CMi   *The Small Dog*

The Small Dog, placed in the sky for his faithfulness, drinks from the Milky Way, once thought to be a river. In Egypt he was Anubis, the jackal god. The Greeks called the whole group Procyon.

α is **Procyon,** meaning "rising before the dog"—that is, before Sirius, so important to the Egyptians. At 11 lt-yr, Procyon is the 14th-nearest star. It is a double. The primary is of type F5IV, magnitude 0.35, with a mass 1.76 times the Sun's. The 11th-magnitude white-dwarf companion, 4" away, 65% as massive as the Sun, orbits the primary in 40.6 years.

β is **Gomeisa,** from the old Arabic constellation name. A B7V star, magnitude 2.91, it is 210 lt-yr away.

CAPRICORNUS

(kăp'-rǐ-kôr'-nŭs)   Capricorni   Cap   *The Sea Goat*

This group, the eleventh constellation of the Zodiac, was recognized by the Babylonians. It lies in the "sea" of the sky along with other "watery" constellations. One Greek legend says that the god Pan, to escape the giant Typhon, leaped into the Nile, and in mid-leap his head, still above water, became the head of a goat, while his hindquarters became the rear part of a fish. Thus was created the Sea Goat. In another legend, Capricornus is the gate through which men's souls pass after death on their way to heaven. In the time of Hipparchus, the Sun in mid-winter was in Capricornus; thus Hipparchus gave the name tropic of Capricorn to the parallel of latitude at the southern limit of the Sun's annual path. Modern eyes have difficulty in seeing a half-goat, half-fish here, but many will recognize the bottom part of a bikini.

α is **Giedi,** "goat." It appears as double, but the stars are unrelated gravitationally. $α_1$, at 1,100 lt-yr, is a cG5 star, magnitude 4.55. $α_2$ is a gG8 star, 116 lt-yr away, magnitude 3.77.

β is **Dabih,** "slaughterers," referring to sacrifices by ancient Arabs at the heliacal rising of Capricornus. This is a double, the 3.06-magnitude primary being of type gK0, the 6th-magnitude companion a late-B. The separation is 205". The system is 130 lt-yr distant.

γ is **Nashira,** "bringing good tidings." It is a dF2 star, 109 lt-yr off, magnitude 3.80.

δ, **Deneb Algedi,** "tail of the goat," is a slightly variable, A6, 3rd-magnitude star at 50 lt-yr.

M30 is an 8th-magnitude globular cluster—for a small telescope.

CANIS MINOR

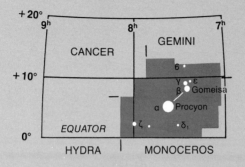

CAPRICORNUS

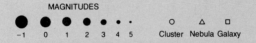

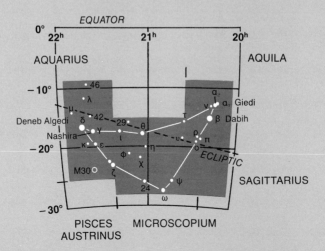

CARINA

(ka-rī'-na)   Carinae   Car   *The Keel*

Argo Navis was built by Argos for Jason and his Argonauts. Later the goddess Athena placed the ship in the sky. It is now divided into Carina, the Keel; Vela, the Sail; Puppis, the Poop or Stern; Pyxis, the Ship's Compass. Greek-letter designations in each part are those assigned before the division. For southern observers, Carina is easily found near Canopus, second-brightest star in the sky. From mid-northern latitudes Carina is not visible.

α, Canopus, was named for the chief pilot of Menelaus on his return from Troy. Being bright and far out of the plane of Earth's orbit, it is often used by spacecraft for navigation. Of magnitude −0.73, type F0Ib, it is 196 lt-yr from Earth.

β, Miaplacidus, is an A0III star, magnitude 1.68, 85 lt-yr off.

ι is **Tureis,** referring to the ornament of a ship's stern. It is of magnitude 2.24, spectral type F0Ib, and 650 lt-yr away.

CASSIOPEIA

(kaš'-ī-ō-pē'-ya)   Cassiopei   Cas   *The Queen*

This easily recognized constellation, like a W or M, appears on 4,000-year-old seals from the Euphrates Valley. In Greek mythology, Cassiopeia was the queen of ancient Aethiopia. She hangs in her chair head downward half of the time—as punishment for the boasting that got Andromeda into trouble (p. 112). Cassiopeia is circumpolar as far south as latitude 50°N. The region is rich in objects for binoculars or a small telescope.

α, Schedar, "the breast," is a K0II star, magnitude 2.22, 147 lt-yr from our solar system.

β is **Caph,** "hand," which early Arabs thought the constellation resembled. An F2IV star, at 45 lt-yr, it varies near 2.26 magnitude.

γ, Cih, is a B0IV star at 96 lt-yr with a variable magnitude around 2.5. It has a 9th-magnitude companion 2″ away.

δ is **Ruchbah,** type A5V, magnitude 2.67, 43 lt-yr away. It may be an eclipsing binary.

ε is a B3IV star of magnitude 3.37, 520 lt-yr distant.

η is a double with components of types G0V and K5, 12″ apart, magnitudes 3.44 and 7.18, and masses 94% and 58% that of the Sun. They orbit a common center in 480 years and are 19 lt-yr away.

ι is a good triple star of magnitudes 4, 7, and 8. A small telescope distinguishes two, possibly three.

M52 and **M103** are 7th-magnitude open clusters, fine in a small telescope.

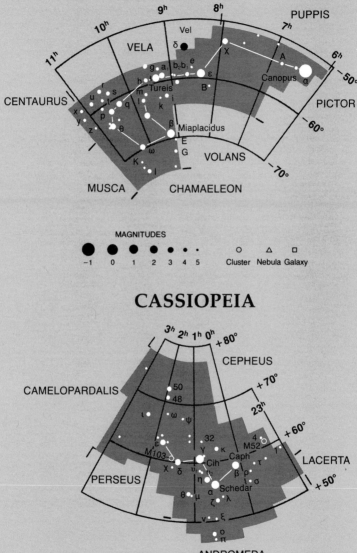

CARINA

PUPPIS

VELA

Vel

δ

CENTAURUS

χ

A

Canopus

α

PICTOR

−50°

−60°

−70°

VEL

g a₁ b₂ b₁ e

h

m l

Tureis

ε

B

s

u

t

q

i

k

x

p

θ

β

Miaplacidus

y

z

ω

E

G

VOLANS

K

MUSCA

CHAMAELEON

MAGNITUDES

●	●	●	●	●	●	·	○	△	□
−1	0	1	2	3	4	5	Cluster	Nebula	Galaxy

CASSIOPEIA

CEPHEUS

CAMELOPARDALIS

50

48

ω ψ

ι

32

M52

4

23ʰ

LACERTA

+80°

+70°

+60°

+50°

ε

γ

κ

Caph

τ

M103

δ

Cih

β ρ

ι σ

PERSEUS

χ

ν₁

η

θ

μ

α

Schedar

ζ λ

ν

ξ

o

π

ANDROMEDA

CENTAURUS

(sĕn-tô'-rŭs)   Centauri   Cen   *The Centaur*

This well-behaved centaur is of Ixion's race, noted mostly as rowdies. He may be Chiron, the centaur skilled in music, though some say Chiron is Sagittarius, the sky's other centaur. The region from Centaurus into Scorpius offers splendid objects for binoculars. In northern latitudes, Centaurus never gets entirely above the southern horizon.

α is **Rigel Kentaurus,** "foot of the centaur," sometimes called Toliman, "grapevine shoot." It is a triple, the closest to the solar system. α Cen A, the primary, is a G2V star, magnitude −0.01, mass 1.08 times that of the Sun. It orbits α Cen B, the secondary, in 80.1 years. α Cen B is a K4 star, magnitude 1.38, and 88% as massive as the Sun. A and B, 22" apart, are orbited by a faint companion of spectral type M5, magnitude 11.05, separation 2.°1. This star is slightly closer to us than the other two and so is called Proxima.

β is **Agena,** "knee" (of the centaur). A close double (separation 1"), it appears as a B1II star, magnitude −0.23, 490 lt-yr away. With α Cen it is spectacular.

ω, once thought to be a single star, is really a globular cluster, 17,000 lt-yr away. To the unaided eye it is a slightly fuzzy "star" of 4th magnitude.

CEPHEUS

(sē'-fūs)   Cephei   Cep   *The King*

"Cepheus" is a two-syllable word, as is "Perseus." Cepheus was the father of Andromeda and husband of Cassiopeia, with whom he ruled ancient Aethiopia (p. 112). Although some people can see a man here, most refer to it as a lopsided house.

α is **Alderamin,** "right arm." This A7IV star of magnitude 2.44 is at a distance of 52 lt-yr. About 5,000 years from now, the north pole of Earth will point close to Alderamin.

β, Alfirk, "the flock," is a B2III star 980 lt-yr away. It is slightly variable around magnitude 3.15.

γ is **Errai,** "the shepherd." At 51 lt-yr this K1IV star has a magnitude of 3.20.

δ is not named but is famous as the prototype of Cepheid variable stars. These have a correlation between the period of variation and the average absolute magnitude; thus they are invaluable as distance indicators. (See pp. 56-57.) δ Cep is 1,300 lt-yr away, and like all other Cepheid variables changes in spectral type as it changes in magnitude. δ goes from F5 to G2 and from 3.51 to 4.42 magnitude during its period of 5.37 days. Cepheids are supergiant stars; δ is 30 times the size of our Sun.

CENTAURUS

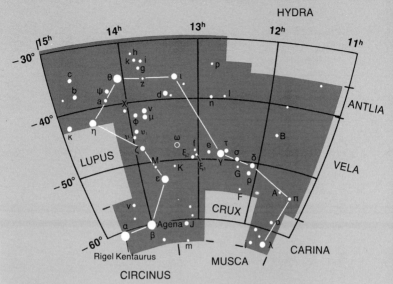

CEPHEUS

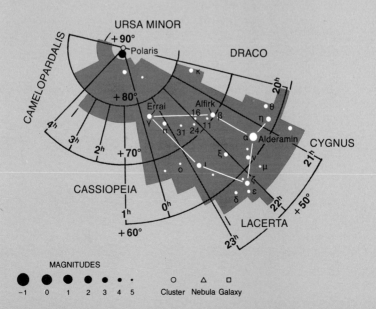

MAGNITUDES

-1 0 1 2 3 4 5   ○ Cluster △ Nebula □ Galaxy

CETUS

(sē'-tŭs)   Ceti   Cet   *The Sea Monster,* or *Whale*

This is the sea monster of the Andromeda legend (p. 112). Modern artists have drawn it as a whale, but this is too tame a creature for the myth. It is an extremely large constellation, identified with the serpent Tiamat of Babylonian mythology. Its brightest stars are only of the 2nd magnitude, but it does have the famous long-period variable star Mira.

α is **Menkar,** "the nose" of the monster. This 2.54-magnitude star, 130 lt-yr from Earth, is of spectral type M2 III.

β, Deneb Kaitos, is "the tail of the whale." This is a K1 III giant star, 59 lt-yr away, with a magnitude of 2.04.

γ is **Kaffaljidhma,** "head of the whale." It is double (separation 3"), appearing as a 3.48-magnitude A2 V star, 68 lt-yr away.

ο is **Mira,** "The Wonderful." The name is from the same root as "miracle." This "LPV" (long-period variable), 130 lt-yr away, changes in brightness from 2.0 to 10.1 magnitude and back again in a period of 332 days. During this cycle it changes in spectral type from M5 to M9. Such stars are very large.

τ, not named, is the 18th-nearest star, only 11.9 lt-yr from the solar system. It is a dG8 star of magnitude 3.50.

CHAMAELEON

(kₐ-mē'-lē-ŭn)   Chamaeleontis   Cha   *The Chameleon*

The Chameleon is an obscure constellation south of Carina, near the other faint southern constellations Musca, Apus, and Volans. Noted by 16th-century navigators, it was first depicted by Bayer in his atlas of 1603 to fill in the far southern sky. It is sometimes depicted as eating Musca the Fly.

α, 67 lt-yr away, is of spectral type F5, magnitude 4.08.

β, 250 lt-yr away, is of spectral type B5 and magnitude 4.38.

γ is a star of spectral type M0, magnitude 4.10, 820 lt-yr distant.

δ is a B5 star of 4.62 magnitude, 408 lt-yr from Earth.

ζ is a B3 star, of magnitude about 5, at a distance not yet determined.

CETUS

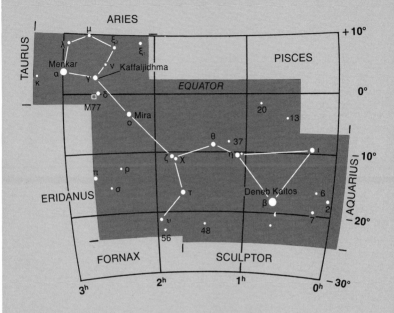

CHAMAELEON

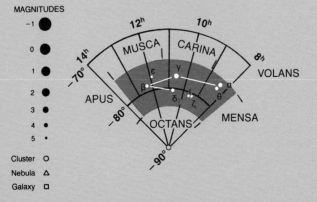

CIRCINUS

(sûr'-sĭ-n*ŭ*s)   Circini   Cir   *The Compass,* or *Dividers*

This is a drawing compass, near the front part of the Centaur. It pivots at the star α with points at β and γ. First figured by Lacaille, this "tool" was to be used with Norma (the Square) and Caelum (the Chisel) by Sculptor. It contains no bright stars.

α, 66 lt-yr away, is a double (separation 16"), appearing as an A8 star of magnitude 3.18. The spectrum shows strong features of strontium.

β is an A3 star of magnitude 4.16. Its distance from Earth has not been determined.

γ is a star of the composite spectral type: B5 and F8. It is 272 lt-yr away, with a magnitude of 4.54.

COLUMBA

(kō-lŭm'-b*a*)   Columbae   Col   *The Dove*

Bayer depicted this constellation to commemorate the dove that Noah sent from the Ark to find dry land. In more ancient times there was a constellation called a dove located in another part of the sky, but the location is unknown today. Columba lies south of Canis Major and Lepus. Some mapmakers thought *Argo Navis* was the Ark, and so depicted Columba sitting on the stern—Puppis.

α is named Phakt, the meaning of which is not known. The name is believed to be modern. This star is 140 lt-yr from Earth. It has a magnitude of 2.64 and is of spectral type B8 V.

β is **Wezn,** which like Wezen (the name of δ Canis Majoris) means "weight." It too is on the poop deck, and was perhaps the weight on the end of a sounding line. The star is of spectral type K2 III, magnitude 3.12, 140 lt-yr from our solar system.

CIRCINUS

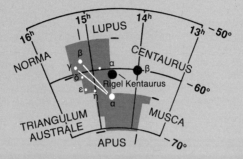

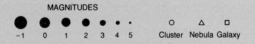

COLUMBA

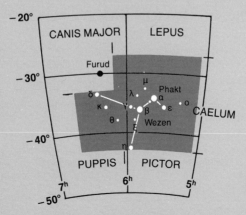

COMA BERENICES

(kō'-m*a* bĕr'-ē-nī'-sēz)   Comae Berenices   Com
Berenice's Hair

This entire constellation consists of an open star cluster. Legends say Berenice, daughter of the king of Cyrene, married Ptolemy Euergetes, a pharaoh of Egypt in the 3rd century B.C. Her hair was famed for its beauty. As an offering of thanksgiving to the gods, she cut it off when her husband returned victorious from battle. The hair was placed in a shrine, but the next night it disappeared. The royal astronomer Conon saved the necks of the priests by claiming the offering had met with such favor from the gods that they had taken it up into the sky. In the same century the Greek astronomer Eratosthenes called this star group "Ariadne's hair." In Tycho's time it became a separate constellation. In this direction lies the galactic north pole, and because the absorbing dust is thin we can see through it and beyond our galaxy into great depths of space.

α is sometimes called Diadem, for a jewel worn in Berenice's hair. It is a dF4 star, 57 lt-yr away, of magnitude 4.32.

β is a dG0 star, 27 lt-yr away and of magnitude 4.32.

γ is a 5th-magnitude K3 star, 300 lt-yr from Earth.

M53 is an 8th-magnitude globular cluster.

M64 is the **Black-eye Nebula**, a 9th-magnitude spiral galaxy 12 million lt-yr off.

The Coma cluster is a group of more than 1,000 galaxies at a distance of some 368 million lt-yr. Several are of 9th or 10th magnitude. Among them:

M83, type S0.

M88, type Sb.

M98, type Sb.

M99, type Sc.

M100, type Sc.

CORONA AUSTRALIS

(kō-rō'-n*a* ôs-trā'-lĭs)   Coronae Australis   CrA
The Southern Crown

This is a not-very-bright semicircle of stars tucked in under Sagittarius. In many legends it was a crown, but Ptolemy called it a wreath, perhaps the laurel wreath worn by champions in the Greek games, and it is thus shown on some charts. Despite the faintness of its stars, it is fairly noticeable owing to its distinctive shape. It has no named stars.

α is an A2 star 102 lt-yr away, of magnitude 4.12.

β is a G5 star, magnitude 4.16, 365 lt-yr away.

γ is a double (separation 2".4) appearing as an F7 star, magnitude 4.26, at a distance of 58 lt-yr.

δ, 220 lt-yr away, is a K1 star of magnitude 4.66.

η is a double star of spectral types A2 and B9. The stars are not gravitationally bound together. Together they are of 6th magnitude.

COMA BERENICES

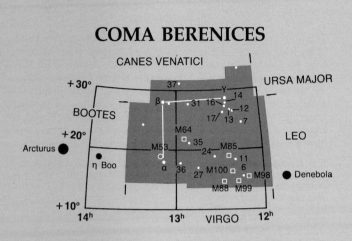

CANES VENATICI

URSA MAJOR

+30°

37

γ 14

β 31 16

ι 12

BOOTES

17 13 7

+20°

M64

35

LEO

Arcturus ●

M53

24

M85

η Boo ●

α 36

27 M100 6 11

M88 M99 M98

● Denebola

+10°

14ʰ   13ʰ   VIRGO   12ʰ

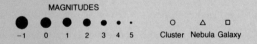

MAGNITUDES

● ● ● ● ● · ·   ○   △   □
-1 0 1 2 3 4 5   Cluster Nebula Galaxy

CORONA AUSTRALIS

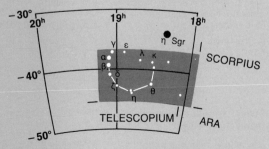

SAGITTARIUS

-30°
20ʰ   19ʰ   18ʰ

γ ε   ● η Sgr

α   λ κ   SCORPIUS

β

-40° δ

ζ

η θ

TELESCOPIUM   ARA

-50°

CRUX

(krŭks)   Crucis   Cru   *The Southern Cross*

This constellation is one of the most beautiful, with its four bright stars in their compact, elegant arrangement. As 16th-century explorers from Europe sailed south of the equator, they found these stars—then included in the large constellation Centaurus—convenient as guides. The upright part of the cross points to the south celestial pole. Crux can be seen up to about latitude 25°N.—about as far north as Key West, Florida. Just to the east of the bottom of the cross is a large dark area, a region of dust, called The Coal Sack.

α is **Acrux** (from its Greek letter name). Near the Centaur's ankle-bone, it is a double star (separation 5″), both components being of type B2IV, 260 lt-yr away. The magnitudes are 1.39 and 1.86.

β is a variable star, magnitude about 1.26, spectral type B0III, at 490 lt-yr.

γ, sometimes called **Gacrux,** is 230 lt-yr away, a giant M3II star of magnitude 1.64.

α is a B2IV star, 570 lt-yr away, slightly variable at magnitude 2.81.

ε is the "misplaced" center star of the cross, a 4th-magnitude K2 star at 172 lt-yr.

CYGNUS

(sĭg′-nŭs)   Cygni   Cyg   *The Swan*

In Roman mythology this is the swan (really, Jupiter in disguise) that wooed Leda. Their offspring were Castor and Pollux, Clytemnestra, and Helen of Troy. For the Arabs, these stars formed an eagle. The whole region is magnificent for scanning with binoculars. The star 61 Cygni was the first to have its distance measured.

α is **Deneb,** "the tail" or top of the Northern Cross. An A2Ia star of magnitude 1.25, it is 1,600 lt-yr away. As our solar system revolves within the Galaxy at 250 km/sec, we are carried toward Deneb.

β is **Albireo,** a beautiful double star distinguishable in binoculars. The components, 35″ apart, are blue and gold, with a combined brightness of 3.07 at 410 lt-yr. The blue star is of type B, the gold one type K3II.

γ is Sadr, "the hen's breast," an F8Ib star of magnitude 2.23 at 815 lt-yr.

σ is B9.5III star, magnitude 2.87, 270 lt-yr away.

ε is **Gienah,** "the wing" (not to be confused with γ Crv, also "Gienah"). It is of 2.46 magnitude, type K0III, 74 lt-yr away.

M29 is an 8th-magnitude open cluster for small telescopes.

M39 is a 6th-magnitude open cluster, good in binoculars.

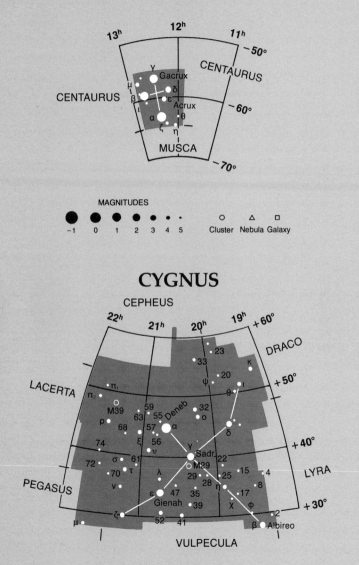

CRUX

13ʰ   12ʰ   11ʰ
−50°

CENTAURUS

γ
Gacrux
δ
CENTAURUS
μ
β   ε
ι
Acrux
θ
α
ζ   η
−60°

MUSCA

−70°

MAGNITUDES

−1   0   1   2   3   4   5

○ Cluster   △ Nebula   □ Galaxy

CYGNUS

CEPHEUS

22ʰ   21ʰ   20ʰ   19ʰ   +60°

DRACO

23
33
κ
20
ψ
θ   ι

LACERTA

π₁
π₂
+50°

M39
59
32
ρ
68
63   55   Deneb
57   α   ο
δ
74
ξ   56
γ
72
σ   61   ν   Sadr   22
70   τ   λ   M29   25   15   4
ν   29   28   η   8
ε   47   35   17
Gienah   39   χ   φ
ζ   52   41   β   2   Albireo

PEGASUS

μ

+40°

+30°

LYRA

VULPECULA

DELPHINUS

(dĕl-fī'-nŭs)   Delphini   Del   *The Dolphin*

This exquisite small star group has been seen in many mythologies as a dolphin—not the fish-dolphin, dorado, but the mammal dolphin, the intelligent and well-loved porpoise. In Greek legend, Delphinus saved the life of the poet-minstrel Arion when he leaped overboard from a ship to escape ruffian-sailors who were threatening his life. The dolphin carried Arion to shore before the ship arrived, and when it did arrive the crew were taken and executed. Some see Delphinus as a small kite—take your choice!

α is **Sualocin,** which along with β, Rotanev, was named in 1814 by astronomers at Italy's Palermo Observatory. They reversed the order of the letters in the name of Nicolaus Venator (a Latin form of Niccolo Cacciatore), assistant to the observatory director. Thus α became Sualocin; and β, Rotanev. α is a B8 star, magnitude 3.86, 270 lt-yr from us.

β, **Rotanev,** is a dF3 star, 96 lt-yr away, magnitude 3.72.

γ, a subgiant K1 star at 112 lt-yr, is a telescopic double (separation 10″.4), with a combined magnitude of about 4.

δ, 250 lt-yr away, is an A5 star of magnitude 4.53.

DORADO

(dō-rä'-dō)   Doradus   Dor   *The Dorado*

Dorado is Spanish for the "fish dolphin" (not the mammal dolphin, or porpoise). This star group was depicted by Bayer in 1603 from reports of sailors in the southern hemisphere. Some observers have incorrectly called Doradus a swordfish. Composed of 3rd- and 4th-magnitude stars, the constellation contains the Large Magellanic Cloud, a satellite galaxy of our Milky Way. The south pole of the ecliptic lies here at right ascension 6ʰ, declination −66°. near the star ξ.

α is an A0III star, 260 lt-yr away, of magnitude 3.28. Its spectrum shows unusual features of silicon.

β is a Cepheid variable, 4.5 to 5.5 magnitude, 910 lt-yr away.

γ, an F5 star of magnitude 4.36, is 47 lt-yr away.

δ is a 4.52-magnitude A5 star at a distance of 148 lt-yr.

LMC is the **Large Magellanic Cloud,** at a distance of 160,000 lt-yr. It is an irregular galaxy, a companion of the Milky Way. In it are many clusters and nebulae, which repay scanning with binoculars or a small telescope. Best is **NGC 2070,** a diffuse nebula located about halfway between the stars ν and θ. Known as the **Great Looped Nebula,** it has also been called the "Tarantula," or "True Lovers' Knot."

DELPHINUS

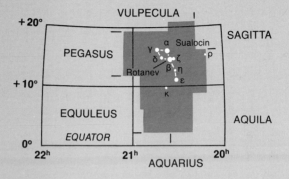

VULPECULA

+20°

SAGITTA

PEGASUS

γ α Sualocin

δ ζ

Rotanev β η ρ

+10°

ε

κ

EQUULEUS

AQUILA

EQUATOR

0°

22ʰ 21ʰ 20ʰ

AQUARIUS

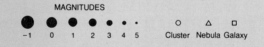

MAGNITUDES

-1 0 1 2 3 4 5 ○ △ □
Cluster Nebula Galaxy

DORADO

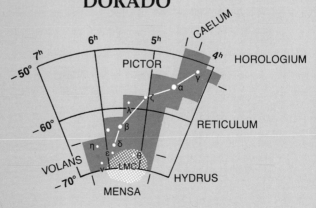

CAELUM

-50°

7ʰ 6ʰ 5ʰ 4ʰ HOROLOGIUM

PICTOR

γ

α

ζ

λ

-60° RETICULUM

β

η δ

ε θ

ν LMC

VOLANS HYDRUS

-70°

MENSA

DRACO

(drā'-kō)   Draconis   Dra   *The Dragon*

Dragons have abounded in mythology for 5,000 years and are useful if you want something guarded. Some say this is the dragon that watched over the golden fleece, but was put to sleep with the potion given to Jason by Medea. Others say that during the war between the gods and the Titans, the goddess Minerva fought a dragon and hurled him into the sky, where he got wrapped around the north celestial pole. Draco can be hard to find simply because he snakes around so much. His tail is between Ursa Major and Ursa Minor, and his head near Lyra and Hercules. Draco is the radiant point for the Draconid meteor showers.

α is **Thuban,** from the Arabic title for this constellation. It is 220 lt-yr away, an A0 star of magnitude 3.64. About 2750 B.C. Thuban, possibly brighter than now, was the north star and as such was worshipped by the Egyptians. Its position has since changed with precession of the equinoxes.

β is **Rastaban,** "the dragon's head." It is a 2.77-magnitude star, a supergiant of type G2II, at 310 lt-yr.

γ is **Eltanin,** also "dragon's head." Brightest star in the constellation, it is of magnitude 2.22, type K5III, at 117 lt-yr.

δ is **Nodus II,** the second "knot," or "loop," of the dragon's body. It is a G9III star, 124 lt-yr away, of 3.06 magnitude.

ζ is **Nodus I,** the first "knot," or "loop." It is a gK3 star, 148 lt-yr away, with a magnitude of 3.22.

λ is **Giansar,** probably meaning "central one." Of type M0III, magnitude 3.84, it is at 188 lt-yr. Look for it halfway between Polaris and the Pointers.

μ is **Arrakis,** meaning "the dancer" for some unknown reason. It is a dF6 star, magnitude 5.06, at 70 lt-yr.

EQUULEUS

(ē-kwoo'-lē-*ŭ*s)   Equulei   Equ   *The Colt*

Equuleus is near Pegasus, the large flying horse. It has little mythology, although noted by ancient writers, including Hipparchus. Probably it was contrived to fill space between Pegasus and Delphinus.

α is sometimes called **Kitalpha,** "part of a horse" (which part not being specified). It is a spectroscopic binary, its spectrum showing features of both dF6 and A3 types. It is 150 lt-yr away and has a magnitude of 4.14.

β is an A2 star 170 lt-yr away, with a magnitude of 5.14.

γ is a close double (2" separation) appearing as an F1 star of magnitude 4.76, at a distance of 180 lt-yr.

δ is a very close double appearing as a dwarf F3 star, magnitude 4.61, 52 lt-yr distant.

ε is a 5.29-magnitude main-sequence F4 star, 170 lt-yr away.

DRACO

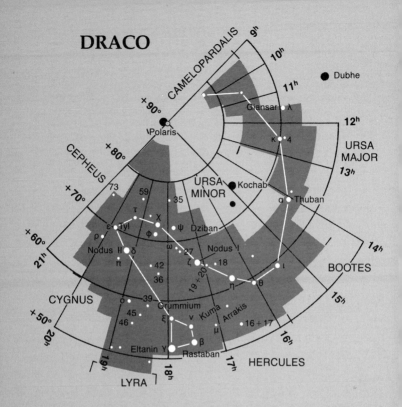

CAMELOPARDALIS

9ʰ
10ʰ
11ʰ
● Dubhe

Giansar λ

+90°

κ 4

12ʰ

Polaris

URSA MAJOR

13ʰ

CEPHEUS

+80°

URSA MINOR

● Kochab

+70°

73

59

35

α Thuban

21ʰ

+60°

τ

χ

ψ

Dziban

Nodus II

ρ

Tyl ε

φ

ω

27

Nodus I

14ʰ

π

δ

42

ζ

18

BOOTES

36

19 +20

η

ι

15ʰ

CYGNUS

+50°

ο

39

Grummium

θ

45

ξ

ν Kuma

Arrakis

20ʰ

46

β

μ

16 + 17

16ʰ

Eltanin γ

Rastaban

17ʰ

HERCULES

19ʰ

18ʰ

LYRA

EQUULEUS

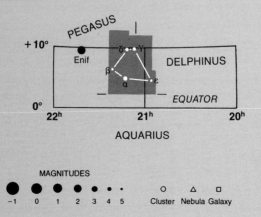

PEGASUS

+10°

δ γ

DELPHINUS

Enif ●

β

α

ε

0°

EQUATOR

22ʰ

21ʰ

20ʰ

AQUARIUS

MAGNITUDES

● ● ● ● ● • ·
-1 0 1 2 3 4 5

○ Cluster △ Nebula □ Galaxy

ERIDANUS

(ē-rĭd′-a-nŭs)   Eridani   Eri   *The River Eridanus*

Almost every culture has identified this winding series of faint stars as a river. It has been identified with major rivers such as the Nile, the Euphrates, and the Po, and also was called the "River of Ocean," said by Homer to encircle a flat Earth. The classical constellation began at the star β, near Orion; it included the stars that are now in Fornax and ended at the star Acamar, from the Arabic for "river's end." Later the constellation was extended to a different star, not visible in mid-northern latitudes: Achernar (another form of Acamar). The stars labeled 53 and 54 were once a separate constellation called the "Brandenburg Scepter."

α, Achernar, "river's end," is one of the brightest stars in the southern sky, at magnitude 0.48. It is a B5 IV star 127 lt-yr from Earth.

β is **Cursa,** "footstool of Orion." This A3 III star is at 78 lt-yr and of magnitude 2.79.

γ is **Zurak,** "bright star of the boat on the river." It is an M0 III giant of magnitude 2.96, and 160 lt-yr away.

δ, Rana, is a 3.72-magnitude dK0 star 29 lt-yr away.

ε, unnamed, is the 9th-closest star to Earth, at a distance of 10.7 lt-yr. It is of spectral type dK2, magnitude 3.73.

32 is a fine 5th- and 6th-magnitude telescopic double star: topaz and blue, respectively.

o² is an interesting triple system, with components of types K1, wdA, and M4, and magnitudes 4.43, 9.53, and 11.17. At 15.9 lt-yr these stars rank 42nd in nearness to Earth. In binoculars **o²** can be seen as a double (separation 82").

FORNAX

(fôr′-năks)   Fornacis   For   *The Furnace*

Originally this was Fornax Chemica, the Chemical Furnace. It was depicted by Lacaille with some faint stars taken from the constellation Eridanus. Fornax contains little of interest except to professional astronomers.

α is a dF5 star, magnitude 3.95, at a distance of 45 lt-yr.

β is a G6 star, 148 lt-yr away, magnitude 4.50.

ν is 470 lt-yr away, a subgiant A0 star of magnitude 4.74.

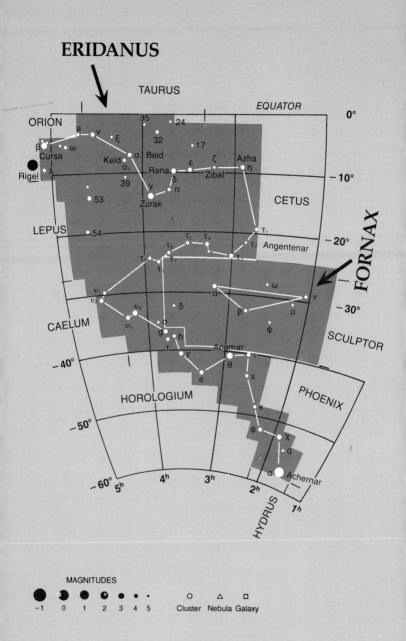

GEMINI

In many mythologies, from the earliest, this star group has been identified with pairs—boy twins, a boy and a girl, Adam and Eve, Hindu deities. It includes the only relatively close pair of nearly identical stars: Castor and Pollux, named for the sons of Leda sired by Jupiter when in his guise as a swan. The brothers, whose sister was Helen of Troy, went with Jason to fetch the golden fleece. The Romans matched their legend with the story of Romulus and Remus. A sky observer with imagination can see the constellation as two stick-figure boys standing side by side. The Sun is in this zodiacal constellation June 21 to July 21. From here radiate the Geminid meteors (p. 228).

α, Castor, appears in a telescope as a close double (separation 2″) of magnitudes 1.97 and 2.95, spectral types A1V and A5. Each is a spectroscopic binary. A third ·binary, 9th magnitude, 73″ away, completes a system of six stars, at a distance of 46 lt-yr.

β, Pollux, magnitude 1.15, is slightly brighter than Castor. It is a K0III star 36 lt-yr from Earth.

γ is **Alhena,** "a mark" (on the feet of Pollux). Of magnitude 1.93 and spectral type A0IV, it is 101 lt-yr from us.

δ, Wasat, is "the middle" (of the constellation). It is a double (separation 7″), appearing as a dA8 star, magnitude 3.51, at 58 lt-yr.

ε, Mebsuta, is of magnitude 3.00, type G8Ib, 1,080 lt-yr away.

η, Propus, marks Castor's left foot. This M3III star, somewhat variable around magnitude 3.33, is 200 lt-yr away.

M35 is a 6th-magnitude open cluster at a distance of 2,800 lt-yr. For binoculars.

GRUS

The Crane was first figured in Bayer's atlas of 1603. This identification seems fitting, because in ancient Egypt the crane was the symbol for a star-watcher. The Arabs, however, had considered these stars part of Piscis Austrinus, the constellation just above Grus.

α, Al Nair, "the bright one," marks the crane's body. Of magnitude 1.74, spectral type B5V, it is 68 lt-yr from Earth.

β is a giant M3II star, 290 lt-yr from us, slightly variable around magnitude 2.17.

γ, the eye of the crane, is 540 lt-yr distant, magnitude 3.00, spectral type B8III.

δ is a red-yellow optical double— not gravitationally connected. The yellow star is a G2, magnitude 4.02, at 233 lt-yr. The red component is an M4, magnitude 4.31, at a distance not determined.

GEMINI

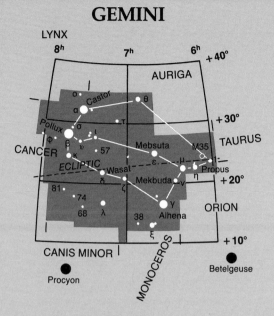

GRUS

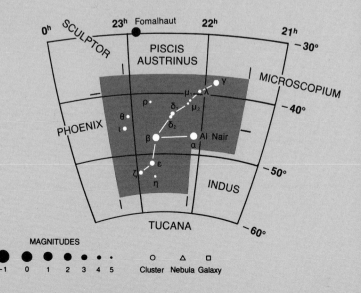

MAGNITUDES

-1 0 1 2 3 4 5

○ Cluster   △ Nebula   □ Galaxy

HERCULES

(hûr'-kū'-lēz)   Herculis   Her   *Hercules, the Strongman*

Recognized from very ancient times as a figure slightly resembling a man, this star group has been associated with innumerable legends. Babylonians thought of it as Gilgamesh, a demigod hero who over-threw the powers of Chaos at the beginning of the world. To the Phoenicians it was the sea god Melkarth. In late classical times it became the Hercules of the Greek legend familiar to us. The "keystone" formed by the stars α, ζ, η, π form the "chest" of Hercules. Some people call the figure marked by η, ζ, β, δ, ε, π, η "the butterfly." The stars of the constellation are not very bright, but it can be located by looking for the keystone about two thirds of the distance from Arcturus (in Boötes) and Vega (in Lyra).

α is **Ras Algethi,** "the kneeler's head." (Most depictions of Hercules show him kneeling.) This is a telescopic double (separation 5"): an M5 II red star of magnitude 3.2 and an F8 star of magnitude 5.39. Because of contrast the F8 star appears green in a telescope. They are 410 lt-yr from us.

β is **Kornephoros,** from an old word meaning "club-bearer." It is a G8 III star of magnitude 2.78 and 103 lt-yr away.

γ, unnamed, is a main-sequence A6 star 141 lt-yr away, with a magnitude of 3.79.

λ is **Maasym,** "the wrist." It is a giant K4 star of magnitude 4.48, at a distance of 233 lt-yr.

M13, a beautiful globular cluster— the most spectacular visible from the northern hemisphere—is visible to the unaided eye as a faint fuzzy "star," although its single brightest star is only 14th magnitude. A 4- to 6-inch telescope can resolve some of the stars, esti-mated to total 100,000 or more. The cluster is 21,000 lt-yr away.

M92 is another globular cluster, about 25,000 lt-yr distant.

ν, ξ, and ο mark the point in the sky toward which the Sun's motion with respect to its neighbors is taking us at a velocity of 19.75 km/sec. The so-called "apex of the Sun's way" is at 18^h04^m and $+30°$.

HOROLOGIUM

(hŏr'-ō-lō'-jǐ-ŭm)   Horologii   Hor   *The Pendulum Clock*

Horologium, a constellation figured in modern times, lies just east of the bright star Achernar (in Eridanus). This obscure group of stars was outlined as a constellation in the mid-eighteenth century by Lacaille. The somewhat crooked pendulum can be discerned, but finding the clock face is a real challenge.

α is a G5 star, 362 lt-yr from Earth, of magnitude 3.36.

β is an A5 star of magnitude 5.08 and undetermined distance.

δ, an F0 star, is 1,100 lt-yr away, of magnitude 4.85.

HERCULES

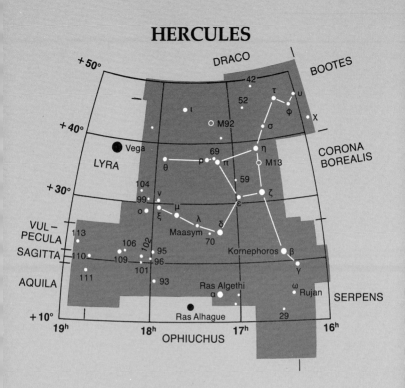

HOROLOGIUM

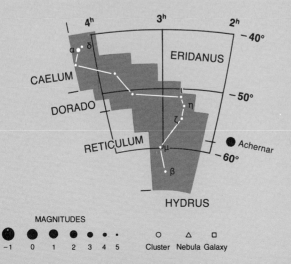

MAGNITUDES

-1 0 1 2 3 4 5

○ Cluster △ Nebula □ Galaxy

HYDRA

(hī'-dra)   Hydrae   Hya   *The Water Serpent*

This star group, the largest of the constellations, should not be confused with another monster—Cetus. Hydra stretches south of the ecliptic from its head, near Cancer, to its tail, near Libra. In Greek mythology, Hydra is the nine-headed serpent of the Lerna marshes, slain by Hercules in his second labor. As one head was cut off, two others grew in its place; but Hercules' faithful Iolaus solved the problem by searing the stump as each head was severed. One head was immortal, and was placed by Hercules beneath a great stone. To the ancient Babylonians these stars were the dragon Tiamat. The Egyptians thought them to be the celestial embodiment of the River Nile. Hydra's brightest star, Alphard, is only of 2nd magnitude; the others are much fainter. To find Hydra, locate Procyon (in Canis Minor) and then follow a long imaginary line southeastward to Centaurus. Alphard is about one fourth of the way down.

α is **Alphard,** "the solitary star in the serpent." In some representations it is in the breast of the serpent. Alphard is a K4 III giant star, magnitude 1.99, 100 lt-yr away.

β, 270 lt-yr away, is a B9-type star with a magnitude of 4.40.

γ is brighter than β, with a magnitude of 2.98, spectral type G8 III, at a distance of 113 lt-yr.

The stars ζ, ε, δ, σ, η, ρ, ζ make up the head of the serpent. This is the immortal head placed under the stone by Hercules.

δ is an A0 star of magnitude 4.18 at 130 lt-yr.

ε is a quadruple star, combined magnitude 3.39, 140 lt-yr from Earth. A small telescope may show the system as a close double (separation 3".6).

σ was at one time called "the snake's nose," but the title is no longer used. This is a giant K3 star of magnitude 4.54, at 272 lt-yr.

M68 is an 8th-magnitude globular cluster at 11.8 lt-yr.

M48 is a 6th-magnitude open cluster, at a distance of 3,100 lt-yr.

M83 is a 7th-magnitude Sc-type galaxy at a distance of 8 million lt-yr from our Milky Way galaxy.

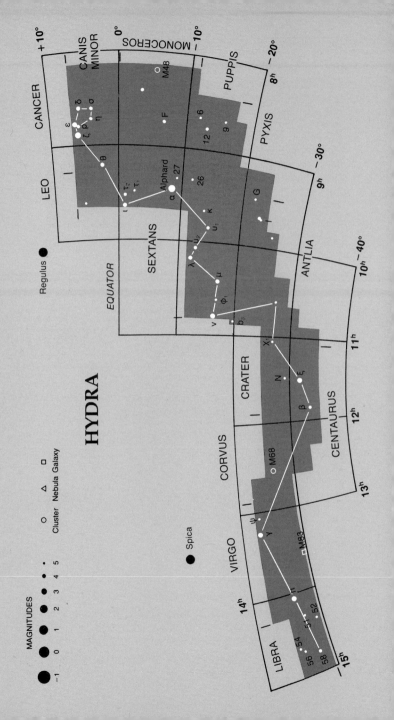

HYDRUS

(hī'-drŭs)   Hydri   Hyi   *The Small Water Snake*

Hydrus (not to be confused with Hydra) is a faint constellation close to the south celestial pole. Its tail reaches due north almost to Achernar (α Eri), and the best way to find it is to find that bright star first. Just outside Hydrus' boundaries to the east are the constellations Mensa and Dorado, which contain the Large Magellanic Cloud. Just outside to the west, in the constellation Tucana, is the Small Magellanic Cloud. Bayer figured Hydrus in his atlas to fill empty space in the southern hemisphere.

α is of magnitude 2.84, 31 lt-yr away, and of spectral type F0 V.

β, a G1 IV star of magnitude 2.78, is 21 lt-yr from Earth.

γ is an M2 giant, of magnitude 3.30, and 300 lt-yr away.

δ, an A2 star of 4.26 magnitude, is at a distance of 72 lt-yr.

ε, a B9 star of magnitude 4.26, lies at a distance of 191 lt-yr.

ζ, an A2 star 652 lt-yr away, is of apparent magnitude 4.90.

η is a double star. The brighter component is a G5 star of magnitude 4.72, at a distance of 125 lt-yr.

ν is a 4.70-magnitude K6 star at a distance of 297 lt-yr.

INDUS

(ĭn'-dŭs)   Indi   Ind   *The (American) Indian*

This is another of the constellations figured by Bayer to fill the southern sky. It contains no bright stars. It is best found by locating the constellation Pavo, the Peacock. Indus is just to the east.

α is a giant K0 star, 84 lt-yr from Earth, of magnitude 3.11.

β is a K2 star of magnitude 3.72 at 270 lt-yr.

δ is 190 lt-yr away, with a magnitude of 4.56 and spectral type F0.

ε is perhaps the most interesting star in this constellation. At a distance of only 11.2 lt-yr, it is the 13th-nearest star to our solar system. It is a K8 main-sequence star of magnitude 4.68.

HYDRUS

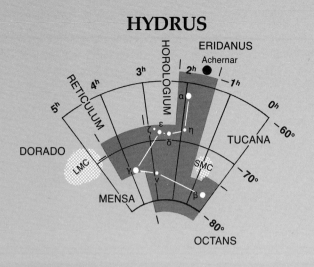

ERIDANUS
Achernar

HOROLOGIUM

RETICULUM

DORADO

TUCANA

MENSA

OCTANS

5ʰ 4ʰ 3ʰ 2ʰ 1ʰ 0ʰ

60°

70°

80°

α ε ζ η δ γ ν β

LMC

SMC

MAGNITUDES

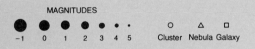

−1   0   1   2   3   4   5   ○   △   ☐
                                Cluster Nebula Galaxy

INDUS

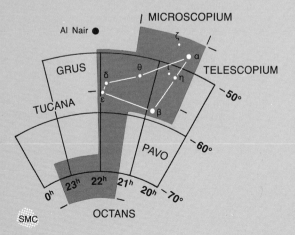

MICROSCOPIUM

Al Nair

GRUS

TUCANA

TELESCOPIUM

PAVO

OCTANS

SMC

ζ α δ θ ι η ε β

0ʰ 23ʰ 22ʰ 21ʰ 20ʰ

50°

60°

70°

LACERTA

(la-sûr'-ta)   Lacertae   Lac   *The Lizard*

For ancient Chinese astronomers, this relatively faint group formed part of a Flying Serpent (the rest consisting of the group we call Cygnus). Lacerta was formed by Hevelius to fill space between Cygnus and Andromeda. For Lacerta he used also the name Stellio (from Latin *stella*, "star"), referring to a kind of Mediterranean newt with starlike dorsal spots. Lacerta has no notably bright stars.

α is a magnitude-3.85 star of type A0, at 91 lt-yr.

β, a gK0 star of magnitude 4.58, is at a distance of 172 lt-yr.

1 Lac is a 4.22-magnitude star of type gK4, 325 lt-yr off.

4 Lac is a cB8 star of magnitude 4.64, at 1,100 lt-yr.

5 Lac, a cK6 star 1,100 lt-yr away, is of magnitude 4.61.

LEO

(lē'-ō)   Leonis   Leo   *The Lion*

The Lion is one of the longest-recognized of the constellations. Ancient Persians, Turks, Syrians, Hebrews, and Babylonians all saw a lion here. In Greece this was the Nemaean lion slain by Hercules in the first of his labors. The Egyptians worshiped these stars because the Sun was in this part of the sky in summer at the time of the annual lifegiving Nile flood. Many observers look not for a lion but for the famous Sickle and the triangle that forms the Lion's hindquarters. The Sun is in Leo from August 12 to September 17.

α is **Regulus,** "little king," so named by Copernicus. It is the heart of the lion—the "dot" of the reversed question mark (otherwise known as the Sickle). Regulus is a B7 V main-sequence star, magnitude 1.35, 85 lt-yr away.

β is **Denebola,** "tail of the lion," an A3 V star 42 lt-yr away, magnitude 2.14.

γ, **Algieba,** is "the lion's mane"— a spectacular double star, separated by 4″, with a combined magnitude of 1.99, at 90 lt-yr. The stars, types G5 and K0 III, appear yellow and green in a small telescope.

δ, **Zosma,** is "the girdle" of the lion. It is a 2.57-magnitude star, type A4V, at 82 lt-yr.

ζ, **Adhafera,** is a giant F0 star of magnitude 3.46 at 130 lt-yr.

θ, **Coxa,** "the hip," is an A2 star, magnitude 3.34, 90 lt-yr distant.

λ, **Alterf,** is a gK5 star 180 lt-yr away, magnitude 4.48.

μ, **Rasales,** "the eyebrows," is a star of magnitude 4.10, type gK3, at 155 lt-yr.

M65, M66, M95, and **M96** are 9th- and 10th-magnitude galaxies, all spirals, visible through a small telescope.

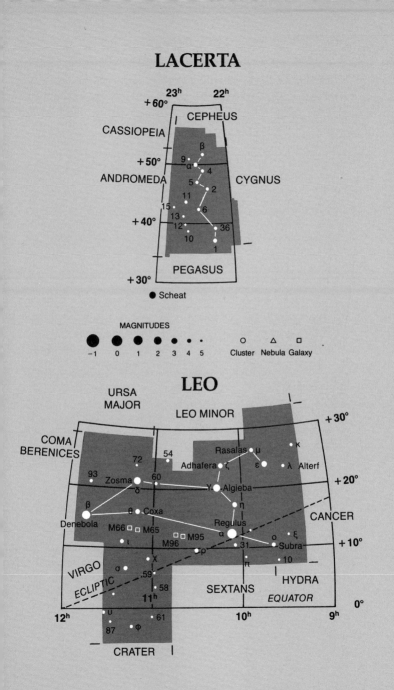

LACERTA

23ʰ   22ʰ
+60°

CEPHEUS

CASSIOPEIA

+50°

β

9
α

4

ANDROMEDA

5
2

CYGNUS

11

15
13
12
10

6

36

1

+40°

+30°

PEGASUS

● Scheat

MAGNITUDES

-1   0   1   2   3   4   5   ○   △   □
                                    Cluster Nebula Galaxy

LEO

URSA
MAJOR

LEO MINOR

+30°

COMA
BERENICES

54

Rasalas μ
Adhafera ζ
ε

κ
λ Alterf

72

93   Zosma
δ
60

γ Algieba

+20°

β
θ Coxa

η

Regulus

CANCER

Denebola

M66 M65

ι

M96 M95

α
31

ξ
ο
Subra

+10°

VIRGO

χ

ρ

π

10

σ
.59

58

SEXTANS

HYDRA

ECLIPTIC

11ʰ

EQUATOR

0°

12ʰ

υ

87 φ

61

10ʰ

9ʰ

CRATER

LEO MINOR

(lē'-ō mī'-nēr)   Leonis Minoris   LMi   *The Small Lion*

This small constellation lies just under the hindfeet of the Great Bear and just over the sickle of Leo. In Arab folklore it was a gazelle with her young; in China, part of a dragon or chariot. In modern times it was named by Hevelius. Leo Minor has no very bright stars.

β is a gG8 star of magnitude 4.41, at a distance of 190 lt-yr. It is not the brightest star in this constellation—46 LMi is. There is no α star.

46 LMi is sometimes called o, the only other Greek letter assigned in this constellation. This 3.92-magnitude sgK2 star is 102 lt-yr away.

10 LMi is a gG6 star of magnitude 4.62 at a distance of 172 lt-yr.

21 LMi is a 4.47-magnitude A5 star, 120 lt-yr distant.

LEPUS

(lē'-pŭs)   Leporis   Lep   *The Hare*

Since the hare was one of Orion the Hunter's favored animals for the hunt, it is placed in the sky just beneath Orion's constellation. Some folklores saw this as the "rabbit in the Moon." To the early Egyptians the stars α, μ, ε, β were the "Boat of Osiris," the god with whom Orion was identified. The Arabs thought of it as four camels drinking from the River Eridanus.

α is **Arneb,** from the Arabic for "rabbit." This is an FO Ib star, at a distance of 900 lt-yr, magnitude 2.58.

β is **Nihal,** Arabic for "camels drinking." This is a double, the primary being of magnitude 2.81, spectral type G5 III, at a distance of 113 lt-yr. The 9.4-magnitude companion is 3" away.

ε is a K5 III star of magnitude 3.21 at 170 lt-yr.

μ is a B9 giant star, magnitude 3.29, at 390 lt-yr.

R Leporis, not marked on the chart, lies due west of the star μ, on the border of Eridanus. R Lep is called "Hind's Crimson Star" for its discoverer and its magnificent red color. It is a variable, ranging from 6th to 11th magnitude in a period of 436 days. R Lep can be found with binoculars when near maximum brightness.

M79 is a 7th-magnitude globular cluster, good for a small telescope.

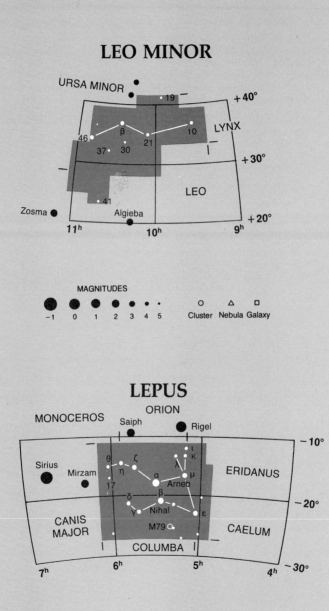

LEO MINOR

URSA MINOR

19

LYNX

46   β   10

37 30 21

+ 40°

+ 30°

LEO

Zosma

41

Algieba

11ʰ   10ʰ   9ʰ

+ 20°

MAGNITUDES

-1 0 1 2 3 4 5   ○   △   ☐

Cluster   Nebula   Galaxy

LEPUS

ORION

MONOCEROS   Saiph   Rigel

- 10°

Sirius   θ   ζ   ι κ

η   λ μ   ERIDANUS

Mirzam   α   Arneb

17   δ   β

γ   Nihal   ε

- 20°

CANIS
MAJOR   M79 ○

CAELUM

COLUMBA

7ʰ   6ʰ   5ʰ   4ʰ

- 30°

LIBRA

(lī'-brα)   Librae   Lib   *The Scales*

As the names of the stars show, for the Greeks this constellation used to be the claws of the Scorpion. Later added by the Romans to the zodiac, this star group is said to have been called The Scales because in classical times the autumnal equinox (when days and nights are "in balance") occurred. Today the Sun is here from November 1 to November 23. Libra is next to Virgo, who is sometimes pictured as Astraea, the goddess of justice, holding The Scales.

α, Zubenelgenubi, "the southern claw of the scorpion," is a double star, the primary having a brightness of 2.76, spectral type A3. The secondary, with a separation of 231", is of magnitude 5.15. The system is at a distance of 66 lt-yr.

β, Zubeneschemali, is "the northern claw of the scorpion," a 2.61-magnitude star of spectral type B8 V, 140 lt-yr away.

γ is Zubenelakrab, "scorpion's claw," a giant G6 star of magnitude 4.02 at 109 lt-yr.

δ is an Algol-type semi-detached binary system. One star orbits the other so closely that they almost touch. They are of combined magnitude 4.8, 100 lt-yr away, of types A0 and dG2, 2.6 and 1.1 solar masses, respectively, and each is 3.5 times the size of the Sun. The orbital period is 2.33 days.

LUPUS

(lū'-pŭs)   Lupi   Lup   *The Wolf*

Classical Greeks and Romans generally knew this star group simply as a wild beast, of a kind unspecified. Eratosthenes saw it as something totally different: a wineskin, held by Centaurus. One classical legend says this was the wolf into which Lycaon was turned by Zeus for the crime of serving human flesh at a banquet for the gods. The Arabs saw in this area of the sky a leopard or a panther.

α is a variable with an average magnitude of 2.32, spectral type B1 V, 430 lt-yr away.

β is a B2 IV star of magnitude 2.69 at 540 lt-yr.

γ is a close double (separation 1") appearing as a B2 V star, 2.80 magnitude, at 570 lt-yr.

δ is 680 lt-yr away, of spectral type B2 IV and magnitude variable around 3.21.

ε is a B3 star, 540 lt-yr off, magnitude 3.74.

φ₁ is a K5 star 272 lt-yr distant, of magnitude 3.59.

φ₂ is a B3 star, 816 lt-yr away, of magnitude 4.69.

LIBRA

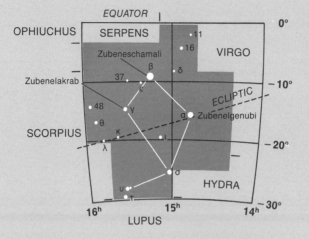

EQUATOR

OPHIUCHUS   SERPENS

Zubeneschamali

Zubenelakrab

SCORPIUS

0°

11
16
VIRGO

β

37
δ

ECLIPTIC — 10°

48
γ
θ
α Zubenelgenubi
κ
ι
λ — 20°

σ
υ
τ
HYDRA
— 30°

16ʰ   15ʰ   14ʰ

LUPUS

LUPUS

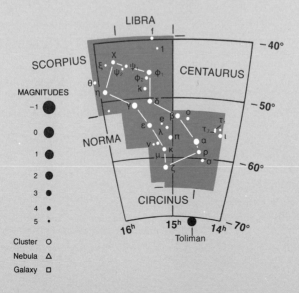

LIBRA
f
1
— 40°

SCORPIUS
χ
ξ
ψ₁
ψ₂
φ₂
φ₁
CENTAURUS

θ
η
κ
— 50°

γ
δ
ο
τ₁

NORMA
ε
β
e
λ
π
τ₂
α
ι

ν
μ
κ
ρ
σ
— 60°

ζ

CIRCINUS

16ʰ   15ʰ   14ʰ — 70°

Toliman

LYNX

(lĭngks)   Lyncis   Lyn   *The Lynx*

These faint stars, below Ursa Major, were depicted as a constellation by Hevelius in 1690. Since the area contains no bright stars, those who wish to see the constellation must be lynx-eyed.

α is an M0 III giant star of magnitude 3.17, 180 lt-yr from Earth.

31 Lyn is a gK5 star, magnitude 4.43, at 217 lt-yr.

38 Lyn is double (separation 2".9) appearing as a B9 star of 3.82 magnitude, 109 lt-yr away.

LYRA

(lī'-ra)   Lyrae   Lyr   *The Lyre,* or *Harp*

In this small, beautiful constellation the six brightest stars form a parallelogram with a triangle at one corner. Vega, third-brightest star in the sky, forms one corner of the triangle. Ancients in India saw here an eagle or a vulture. Early Greeks saw it as their lyre, often with the body formed from a tortoise shell (from which, some thought, the first lyre was contrived). Old Greek legend claims this is the lyre of Orpheus, musician of the Argonauts and husband of Eurydice, and that the lyre was placed in the sky to commemorate the beautiful music it produced at the hands of Orpheus.

α is **Vega,** or **Wega,** "eagle" in Arabic. Spectacularly bright, it passes high over cities near lat. 40°N. Blue-white, of magnitude 0.04 and spectral type A0 V, it is one of the few bright stars that are not giants or supergiants. Vega is 26 lt-yr away. Our solar system is moving toward it at about 19 km/sec.

β is **Sheliak,** Arabic for "tortoise." This is an eclipsing binary—magnitude 3.38-4.36, period 12.9 days. The components are of spectral types B2 and F, 1,300 lt-yr away. This double shows a peculiar spectrum, for each star is throwing off a stream of gas which spirals around it. The American astronomer Otto Struve spent so much of his career studying this system that it is sometimes jokingly referred to as "Beta Struve."

γ is **Sulaphat,** another Arabic name for "tortoise." It is a B9 III star of magnitude 3.25, at 370 lt-yr.

ε is a famous "double double." In binoculars (and to a very sharp eye) it is a pair, ε¹ and ε², of magnitudes 5.1 and 4.4, respectively, 208" apart. A good small telescope can split each—ε¹ into stars of 5.4 and 6.5 magnitude, 2".7 apart, with an orbital period of 1,200 years, and ε² into stars of 5.1 and 5.3 magnitude, 2".3 apart, with an orbital period of 600 years.

M57 is the famous "Ring Nebula," a planetary nebula 5,000 lt-yr away. A small telescope is required to see it, since it is 9th magnitude. The central star which threw off this sphere of gas is 15th magnitude, at the center of the ring. A large telescope is required to see it.

LYNX

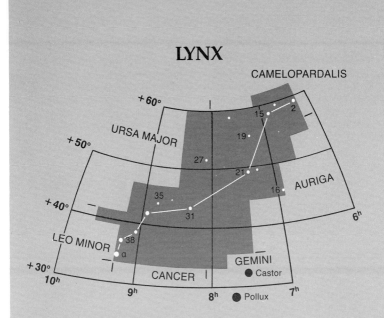

LYRA

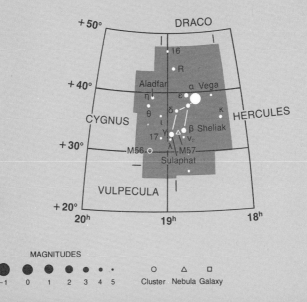

MAGNITUDES

MENSA

(mĕn'-sa)   Mensae   Men   *The Table Mountain*

This was originally named Mons Mensae, the "table mountain," by La-caille, referring to Table Mountain (often cloud-capped) above Cape Town, South Africa. He placed it near the south celestial pole, next to the obscure constellation Octans. Later the "Mons" was dropped, so that sometimes this constellation is called merely "the table." It contains no bright stars, but at its border is the Large Magellanic Cloud.

α is a dG6 star of magnitude 5.14, at a distance of 28 lt-yr.

β is in the Tarantula Nebula. The star has a magnitude of 5.30 and is of spectral type G8. It is 136 lt-yr away.

γ is 220 lt-yr away, of spectral type K4 and magnitude 5.06.

η, of magnitude 5.28, is a type K6 star at a distance of 192 lt-yr.

μ is a B9 star of magnitude 5.69—distance undetermined.

MICROSCOPIUM

(mī'-krō-skō'-pī-ŭm)   Microscopii   Mic   *The Microscope*

Lacaille figured this faint constellation south of Capricornus and east of Sagittarius. Composed of 5th-magnitude stars, it was intended to complement the new constellation Telescopium, nearby. These constellations were originated by Lacaille to commemorate the exploration of the microcosm and the macrocosm—the smallest and the largest aspects of the universe.

α is a star of 5.00 magnitude, spectral type G6. It is 365 lt-yr from our solar system.

γ is a G4 star, 230 lt-yr away, of magnitude 4.71. At one time it was part of Piscis Austrinus.

ε is an A0 star, 112 lt-yr away, of magnitude 4.79. It, too, was once part of Piscis Austrinus.

MENSA

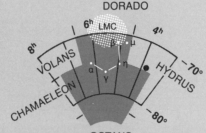

DORADO

6ʰ LMC

β γ μ

8ʰ

VOLANS

4ʰ

−70°

η

α

HYDRUS

CHAMAELEON

γ

−80°

OCTANS

MAGNITUDES

-1   0   1   2   3   4   5   ○   △   ☐

Cluster   Nebula   Galaxy

MICROSCOPIUM

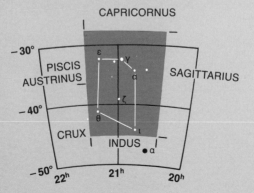

CAPRICORNUS

−30°

ε

γ

PISCIS

AUSTRINUS

α

SAGITTARIUS

−40°

ζ

CRUX

θ

ι

INDUS

● α

−50°

22ʰ   21ʰ   20ʰ

MONOCEROS

(mō-nŏs'-ēr-ŏs)   Monocerotis   Mon   *The Unicorn*

This constellation, figured probably by Bartsch, straddles both the celestial equator, just east of Orion, and the galactic equator, which passes close to the stars 13, 18, and 19, cutting the celestial equator at a 67-degree angle. It contains no very bright stars.

α is a star of magnitude 4.07 and spectral type gK0. It is 180 lt-yr from Earth.

β is a telescopic triple star, with 5th-magnitude components separated by 7″ and 10″. Two of the stars are of type B3 and are 470 lt-yr away.

γ is a star of magnitude 4.09 and spectral type K2. It is 250 lt-yr from the solar system.

δ is a 4.09-magnitude A0-type star at a distance of 180 lt-yr.

15 Mon is a variable star (about 4.2 to 4.7 magnitude) known also by the designation S Mon. Of spectral type O7, it is 408 lt-yr away.

M50 is a 7th-magnitude open cluster for a small telescope.

NGC 2244 is visible to the unaided eye as a beautiful open cluster in the form of a rosette. Relatively young as stars go, it is 5,300 lt-yr away. For better viewing, try binoculars.

MUSCA

(mŭs'-ka)   Muscae   Mus   *The Fly*

Bayer depicted this inconspicuous star group as a bee. Later Lacaille depicted it as a fly. It was said to be a match for the Northern Fly, Musca Borealis, once placed on the back of Aries, but now not recognized as a constellation. Musca lies just south of Crux, the Southern Cross.

α is a slightly variable star of magnitude 2.66 to 2.73 and spectral type B2 IV. It is at a distance of 430 lt-yr.

β is a double star whose components are of magnitudes 3.7 and 4.0, shining with a combined light of magnitude 3.06. They appear as a B2V star, 470 lt-yr away. Since the components are separated by only 1″, very good seeing is required to split the pair with a small telescope.

δ is a K2 star of magnitude 3.63 at a distance of 155 lt-yr.

MONOCEROS

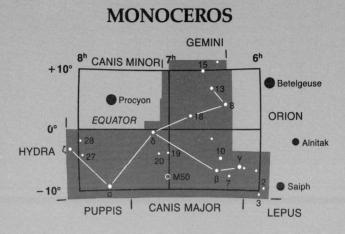

GEMINI

+10°   8ʰ CANIS MINOR | 7ʰ   6ʰ
                                        Betelgeuse
                 Procyon   13
         EQUATOR   18   8   ORION
0°   Alnitak
HYDRA ζ 28   10   γ
          27   δ 19 20   β   7
-10°   α   M50   2 Saiph
                                               3
       PUPPIS | CANIS MAJOR | LEPUS

MAGNITUDES

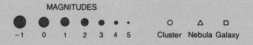

-1   0   1   2   3   4   5   ○   △   □
                                 Cluster Nebula   Galaxy

MUSCA

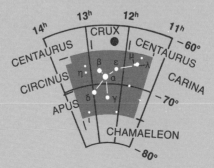

        13ʰ   12ʰ
   14ʰ   CRUX   11ʰ
                        -60°
CENTAURUS   μ CENTAURUS
              β ε   λ
CIRCINUS   η α   CARINA
              δ   γ
APUS   -70°
   ι
        CHAMAELEON
                -80°

NORMA

(nôr'-ma)   Normae   Nor   *The Carpenter's Square*

Originally this was Norma et Regula, the Latin for "carpenter's square and level." (From Latin *norma*, "rule," we get our word "normal.") These tools, along with nearby Caelum and Circinus, were for use in the Sculptor's workshop. Located north of, and adjoining, Triangulum, Norma was formed by taking some stars from Ara and Lupus. The region is very rich in celestial objects for binoculars or a low-power telescope.

δ is a 4.84-magnitude star of spectral type A3, at a distance of 230 lt-yr.

ε is a B5 star of magnitude 4.80, at 650 lt-yr.

η, a 4.74-magnitude star of spectral type G4, is 190 lt-yr away.

γ₁ is a supergiant G4 star of magnitude 5.00, distance not determined.

γ₂ is a 4.14-magnitude star of type G8, 82 lt-yr away.

OCTANS

(ŏk'-tănz)   Octantis   Oct   *The Octant*

This star group was originally called Octans Hadleianus (Latin for "Hadley's Octant") to commemorate the invention of this important navigational instrument by John Hadley in 1730. The constellation was first listed by Lacaille in 1752. An octant is like a sextant, but its measuring arc is one eighth of a circle, whereas the arc of a sextant is one sixth of a circle. Since the mirrors on these instruments double the angles that can be measured, an octant can measure an angle (usually the altitude of a star above the horizon) up to 90°. The sextant, invented later, enabled navigators to measure angles up to 120°; so octants are rarely used today. Octans is devoid of bright stars but is appropriately placed as a navigation aid, since it includes the south celestial pole. Unlike the north celestial pole, marked approximately by Polaris, the south celestial pole is unmarked by any bright star. The nearest star visible to the naked eye is σ, of 5th magnitude, 52' away from the pole. Thus, the pole is difficult to locate exactly.

α is a star of magnitude 5.24 and spectral type F4. It is 148 lt-yr from Earth.

β is an F1 star of magnitude 4.34, distance not determined.

δ is a K2 star of magnitude 4.14, at 205 lt-yr.

θ is a K5 star of magnitude 4.73 at 250 lt-yr.

σ, the star closest to the south celestial pole (for the unaided eye), is an A7 star of magnitude 5.48, distance not determined.

NORMA

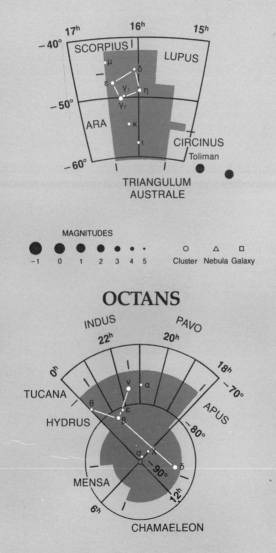

17ʰ 16ʰ 15ʰ

−40°

SCORPIUS LUPUS

μ
δ
ε
γ₁ η
γ₂

−50°

ARA

κ

ι

CIRCINUS
Toliman

−60°

TRIANGULUM
AUSTRALE

MAGNITUDES

-1 0 1 2 3 4 5   O   △   □
                        Cluster Nebula Galaxy

OCTANS

INDUS   PAVO

22ʰ   20ʰ

0ʰ   18ʰ

−70°

TUCANA

ν   α

θ
ε

HYDRUS

β

APUS

−80°

σ
□   δ
−90°

MENSA

12ʰ

6ʰ

CHAMAELEON

OPHIUCHUS

(ŏf'-ĭ-ū'-kŭs)   Ophiuchi   Oph   *The Serpent Bearer*

In Greek legend the Serpent Bearer is the god Aesculapius, founder of medicine, ship's doctor for the Argonauts. He was so skillful that he brought a dead man back to life. This power so worried Pluto, god of the underworld, that he persuaded Jupiter to place Aesculapius among the stars, out of the way. When the zodiac was formed, in classical times, the Sun did not pass through Ophiuchus, but because of precession the Sun today spends more time in Ophiuchus than in Scorpius. It is in Ophiuchus from November 30, when it leaves Scorpius, until December 18, when it enters Sagittarius.

α is **Ras Alhague,** "head of the serpent charmer," a star of magnitude 2.07, spectral type A5 III, at 60 lt-yr.

β is **Cebalrai,** "heart of the shepherd," a K2 III star of magnitude 2.77, at 124 lt-yr.

η, **Sabik,** is "the preceding one." It is a close double, both components being of type A2, 69 lt-yr away, combined magnitude 2.43, separation 1".

70 Oph, 16.7 lt-yr away, is the 46th-nearest star. It is double, with components of types K0 and K5, combined magnitude of 4, separation of 2".8, and a period of 88 years.

λ is **Marfik,** "the elbow," an A1 star of magnitude 3.85, at a distance of 192 lt-yr.

All M-objects on the chart are globular clusters. **M9** is of 8th magnitude; **M10,** 7th magnitude and 20,000 lt-yr away. **M12** is of 8th magnitude, 24,000 lt-yr away. **M14** is 8th and **M19** 7th magnitude, as is **M62. M107** is of 9th magnitude.

SERPENS

(sûr'-pĕnz)   Serpentis   Ser   *The Serpent*

Serpens, the serpent held by Ophiuchus, is in two parts, at the Serpent Bearer's sides. The Head, Serpens Caput (Latin *caput* means "head," not "dead"!), is to the west; Serpens Cauda, the Tail, to the east.

α is **Unakalhai,** "the serpent's neck," a giant K2 star, magnitude 2.65, 71 lt-yr away.

β is an A0 star of magnitude 3.74 at 120 lt-yr.

μ is of spectral type A0 and magnitude 3.63 at 192 lt-yr.

θ, **Alya,** is a double star with both components of type A5, of 5th magnitude combined, separation 22", 142 lt-yr away.

ξ is an A5 star of magnitude 3.64 at 105 lt-yr.

η is a K0 III star, magnitude 3.23, 60 lt-yr distant.

M5 is a globular cluster, appearing like a star of magnitude 6.7, at 26,000 lt-yr.

M16 is an open cluster enveloped in a nebula of 7th magnitude, 5,500 lt-yr distant.

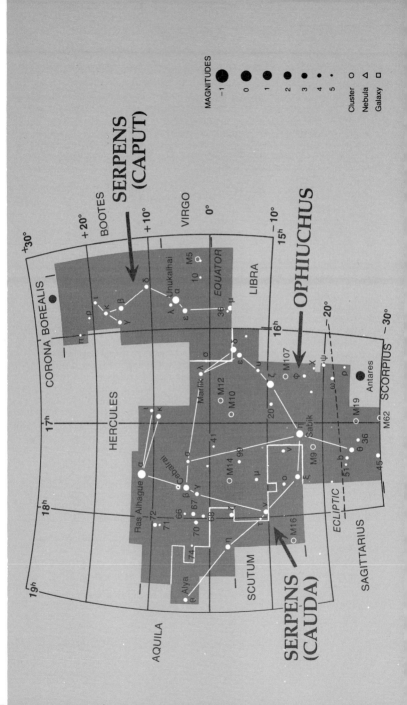

ORION

(ō-rī'-ŏn)   Orionis   Ori   *The Hunter*

This majestic figure does suggest that of a giant. Straddling the celestial equator, it is known all over the world. From Mesopotamian times it has heralded storms and winter. In earliest Egyptian mythology, it marked the resting place of the soul of Osiris. In Greece, Orion as the son of Neptune was the tallest and handsomest of men, Diana's lover, pursuer of the beautiful Pleiades. Killed by a scorpion, he was placed in the sky by Diana far from Scorpius, for his protection. Orion's famous Belt points to Sirius (his dog) and Canis Major in the east, to Aldebaran and Taurus in the west.

α is **Betelgeuse** (bĕt'-el-gĕrz), "armpit of the central one." A giant M2I star, it is somewhat variable about magnitude 0.8, quite red, 652 lt-yr away.

β, Rigel (rī'-jĕl), is the brightest star in Orion. A B8Ia star, magnitude 0.11, it marks the left leg. At 815 lt-yr, it has a 7th-magnitude companion 9″ away.

γ, Bellatrix, the left shoulder, is named for an Amazon. The star is of magnitude 1.63, spectral type B2III, 303 lt-yr away.

κ is **Saiph,** "the sword" (hardly appropriate for its position in Orion's right foot!). It is a B0.5I star, 1,826 lt-yr away, of magnitude 2.05.

δ, Mintaka, very near the celestial equator, is the top star in the Belt. An eclipsing variable with a period of 5.7 days, it is of type O9.5II, magnitude 2.19, 1,500 lt-yr away.

ε, Alnilam, "string of pearls," is the middle star of the Belt. It is a B0Ia star, 1,532 lt-yr away, magnitude 1.70.

ζ, Alnitak, "the girdle," is the southernmost star of the Belt. It is a double (3″ separation) of combined magnitude 1.79, combined spectral type O9.5Ib, at 1,467 lt-yr.

M42 is the Great Nebula in Orion, one of the showpieces of the sky. To the unaided eye this diffuse nebula appears as a slightly fuzzy, greenish "star." A telescope reveals the vast, swirling clouds of gas, in the midst of which are the stars ι and θ Ori. The latter is quadruple—the so-called Trapezium, all very young stars of 6th to 8th magnitude. Nearby is the smaller diffuse nebula **M43;** another is **M78.**

PAVO

(pā'-vō)   Pavonis   Pav   *The Peacock*

This constellation was depicted by Bayer. In Greek myth the stars that are now the Peacock were Argos, builder of the ship Argo. He was changed by the goddess Juno into a peacock and placed in the sky along with his ship. This bird has long been a symbol of immortality.

α, "Peacock," is a B3IV star of magnitude 1.93, 293 lt-yr off.

β is an A7III star, 160 lt-yr away, of magnitude 3.45.

γ is an F8 star of magnitude 4.30, at 29 lt-yr.

172

ORION

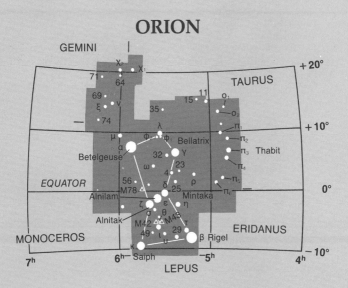

GEMINI

TAURUS

χ₂ χ₁
71
64
69
ξ ν
74
35
15 11
o₁
o₂
π₁
π₂
π₃ Thabit
π₄
π₅
π₆

λ
φ Φ
μ
α
Bellatrix
Betelgeuse
32
γ
ω
23
4
56
δ
ρ
M78
25
Alnilam
ζ
ε
η
Mintaka
Alnitak
θ
σ
M42 M43
49 ι
29
υ
β Rigel
κ
Saiph

EQUATOR

MONOCEROS

ERIDANUS

LEPUS

+20°
+10°
0°
-10°

7ʰ 6ʰ 5ʰ 4ʰ

PAVO

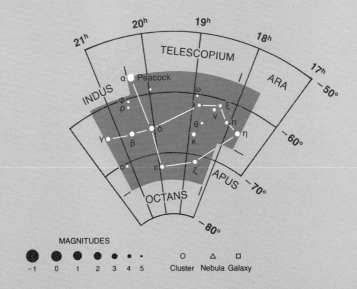

21ʰ 20ʰ 19ʰ 18ʰ 17ʰ

TELESCOPIUM

ARA

INDUS

α Peacock
φ
ρ
ω
λ
ξ
ν
γ
β
δ
θ
π
κ
η
o
ε
ζ

OCTANS

APUS

-50°
-60°
-70°
-80°

MAGNITUDES

● ● ● ● ● • ·
-1 0 1 2 3 4 5

○ △ □
Cluster Nebula Galaxy

PEGASUS

(pĕg'-*a*-sŭs)   Pegasi   Peg   *The Winged Horse*

This star group was seen as a winged horse even in preclassical times. In Greek mythology, when Perseus cut off the head of Medusa some blood fell into the sea, mixed with foam, and formed white-winged Pegasus. This steed with a kick started the fountain Hippocrene on Helicon, mountain of the Muses, and was ridden by the hero Bellerophon when he slew the Chimera. Only the horse's front half is in the sky; the other half, according to modern cynics, fell to earth, there to begin the breed of politicians. The famous Great Square—the stars α, β, and γ Peg, plus α And—is a landmark of the sky.

α, **Markab,** "the saddle," is a B9.5 III star of magnitude 2.50, at 109 lt-yr.

β is **Scheat,** "the shoulder" (of the horse). Of type M2 II, it varies between 2.4 and 2.7 magnitude, at 210 lt-yr.

γ, **Algenib,** is the "wing"' or "side." It is a B2 IV star, slightly variable around magnitude 2.84, at 570 lt-yr.

ε, **Enif,** "the nose," is a K2 Ib star of magnitude 2.38, at 780 lt-yr.

η, **Matar,** "fortunate rain," is of type G8 II, 360 lt-yr away, magnitude 2.95.

ζ, **Homam,** "lucky star," is of type B8 V, magnitude 3.40, 210 lt-yr away.

M15 is a 6th-magnitude globular cluster at 34,000 lt-yr.

PERSEUS

(pûr'-sūs)   Persei   Per   *Perseus, the Hero*

Here is the youth who saved Andromeda from Cetus. He still holds the Gorgon's head. Her eye—Algol, the Demon Star—still winks. Some see this constellation as a large letter "K" or a fleur-de-lis.

α is **Algenib,** "the side," or **Mirfak,** "the elbow," an F5 Ib star of magnitude 1.80 at 522 lt-yr.

β is **Algol,** "the demon," or Demon Star, which "blinks," It changes in brightness from 2.06 to 3.28 magnitude and back again every 2.87 days. This is an eclipsing binary, a B8 and a gK0 star. When the larger (gK0) star moves in front of the other, the brightness diminishes for 10 hours. The pair are 105 lt-yr away.

ξ, **Menkib,** was "the shoulder" but is now the ankle. It is an O7 star of magnitude 4.05, at a distance of 2,100 lt-yr.

h and x are the **Double Cluster,** the former at 7,000 lt-yr, the latter at 8,000. Both are of 9th magnitude—spectacular in binoculars or a small telescope.

M34 is a 6th-magnitude open cluster, for a small telescope.

M76, a planetary nebula, appears as an 11th-magnitude star. A medium-size telescope is required to make much of this object, 15,000 lt-yr away.

PEGASUS

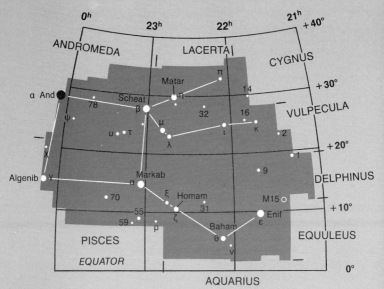

PERSEUS

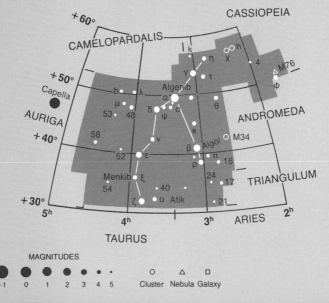

MAGNITUDES

-1 0 1 2 3 4 5

○ Cluster △ Nebula □ Galaxy

PHOENIX

(fē'-nĭks)   Phoenicis   Phe   *The Phoenix*

This star group represents the mythical bird that in ancient Egypt was sacred to the god Ra and a symbol of periodic renewal and immortality. In one story, the bird consumes itself every 500 years on a pyre and rises from the ashes reborn. The Arabs knew it as an ostrich. Bayer was the first to define this constellation, placing it just south of Fornax and Sculptor, west of Achernar.

α is **Ankaa,** a K0 giant star, magnitude 2.39, at a distance of 93 lt-yr.

β is a close double (separation 1") of combined magnitude 3.30 and spectral type G8 III, 190 lt-yr away.

γ, 1,300 lt-yr from our solar system, is a K5 IIb-III star of magnitude 3.40.

δ is a G4 star, 120 lt-yr from us, magnitude 3.96.

ζ is a double (separation 1".3), appearing as a B7 star, 402 lt-yr from Earth, with a magnitude of 4.13. A small telescope may split it.

PICTOR

(pĭk'-tēr)   Pictoris   Pic   *The Painter's Easel*

Originally this star group was Equuleus Pictoris. "Equuleus" can mean ass, small horse, or easel—the latter perhaps from an old custom among artists of carrying a canvas on a donkey. In its present form the name means simply "painter." This faint constellation was formed by Lacaille south of Columba and north of the Large Magellanic Cloud. The easiest way of locating it is to look west of Canopus.

α is an A7 V star, of magnitude 3.27, at a distance of 57 lt-yr.

β, an A3 star, magnitude 3.94, is 62 lt-yr away.

γ is of spectral type K1, at a distance of 240 lt-yr with a magnitude of 4.38.

η is two stars, the brighter component, η_2, being an M2 star of magnitude 4.92, at a distance of 1,800 lt-yr. η_1, the fainter star, widely spaced, is of type F5, magnitude 5.44, at a distance of 72 lt-yr from Earth.

PHOENIX

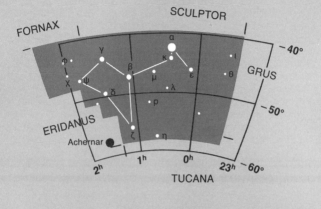

MAGNITUDES

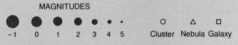

-1 0 1 2 3 4 5   ○   △   □
                          Cluster Nebula Galaxy

PICTOR

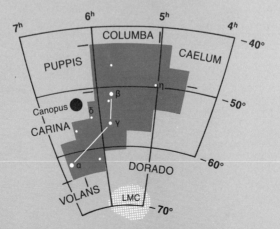

PISCES

(pĭs'-ēz)   Piscium   Psc   *The Fishes*

Folklore has identified this faint star group as fishes since early Mesopotamian times. One Greek story, based on Syrian legend, says the
fishes are Venus and her son Cupid, who escaped Typhon, the fire-
breathing giant, by jumping into the Euphrates River and turning into
fishes. A Roman story relates that the stars here are fishes that carried
Venus and Cupid to safety in the river. The so-called Eastern Fish is
defined by a circlet of stars below the Square of Pegasus; the Western
Fish is to the west of Pegasus and south of Andromeda. The two are
connected by cords tied to their tails and joined at the star α, Alrisha,
"the knot." In this constellation the ecliptic crosses the celestial equator
at the vernal equinox, where the Sun is at the beginning of spring. That
location is still called the First Point in Aries, since it was located in the
Ram 2,000 years ago when the zodiac was defined in its present form.
The Sun is now in Pisces from March 13 to April 19.

α, Alrisha, "the knot," is a double with components of type A2 and magnitudes 4.3 and 5.3, at 130 lt-yr.

β is 325 lt-yr away, magnitude 4.58, spectral type B5.

γ, brightest star of the Western Fish, is a 3.85-magnitude G5 giant, 125 lt-yr off.

η is a giant G3 star of magnitude 3.72, at 450 lt-yr.

ω, a dF3 star, is of magnitude 4.03, at 157 lt-yr.

τ is a sub-giant K1 star of magnitude 4.70, 180 lt-yr away.

M74 is a 9th-magnitude open-armed galaxy, type Sc, at 7 million lt-yr.

PISCIS AUSTRINUS

(pĭs'-ĭs ôs-trī'-nŭs)   Piscis Austrini   PsA   *The Southern Fish*

This faint constellation was the symbol of the old Syrian god Dagon,
but in Greek lore was seen as a fish. Just below Aquarius, it is inconspicuous except for its brightest star, Fomalhaut. In this constellation,
stars are depicted as a fish in the river Eridanus.

α, Fomalhaut, is "the fish's mouth." This star has a magnitude of 1.16, is of spectral type A3V, and is 23 lt-yr away. At its declination (−30°), this bright star skims the southern horizon when observed from mid-north-ern latitudes.

β is an A0 star of magnitude 4.36. It is 220 lt-yr from our solar system.

ι also is an A0 star. It is 103 lt-yr away, with a magnitude of 4.35.

ε, a B8 star 250 lt-yr from us, has a magnitude of 4.22.

PISCES

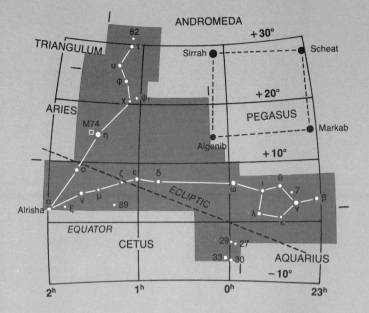

PISCIS AUSTRINUS

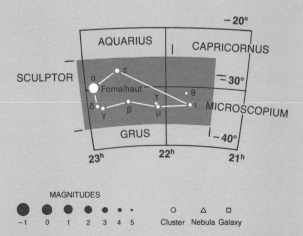

MAGNITUDES

-1 0 1 2 3 4 5   Cluster Nebula Galaxy

PUPPIS

This is the stern of Argo, the Ship. Below Puppis is Carina, the Keel; to the east is Vela, the Sail; and just beside Puppis on the deck is Pyxis, the Ship's Compass. To find these stars, locate Canopus and then look northward for a group of 2nd- and 3rd-magnitude stars. Because Puppis was once part of Argo, the Greek letter designations are not in order of brightness, and some letters are missing.

ρ, an F6II star of magnitude 2.80, is 105 lt-yr from Earth.

ξ, Asmidiske, is a G3Ib star of magnitude 3.34. It is 1,240 lt-yr distant.

ζ, Naos, is "the ship." One of the most intrinsically luminous stars, it is of spectral type O5, magnitude 2.25, and 2,300 lt-yr from the solar system.

π is a gK4 star of magnitude 2.70, at a distance of 140 lt-yr.

σ is a K5III star, of magnitude 3.24, at a distance of 180 lt-yr. It has a 9th-magnitude companion 22" away.

ν is of magnitude 3.19, a B7 giant star 620 lt-yr away.

τ is a K0III star, magnitude 2.92, at a distance of 124 lt-yr.

M46 is a 7th-magnitude open cluster, 5,400 lt-yr away. Good in a small telescope.

M47 is a 5th-magnitude open cluster, 3,750 lt-yr away.

M93 is a 6th-magnitude open cluster, 3,600 lt-yr distant.

PYXIS

Pyxis hardly looks like what its name means: a compass on the bridge of a ship. It was once part of Argo Navis. Pyxis, containing no bright stars, was defined by Lacaille from stars formerly in Vela; thus its Greek-letter star names were chosen in modern times and are unrelated to the names the stars had when part of Argo. To find this group, locate Canopus, work your way through Puppis, and then look just east of the middle section of Puppis.

α is a B1 star of magnitude 3.70. It is 517 lt-yr from us.

β, a G5 star at a distance of 260 lt-yr, is of magnitude 4.04.

γ is a K4 star, 213 lt-yr away, of magnitude 4.19.

PUPPIS

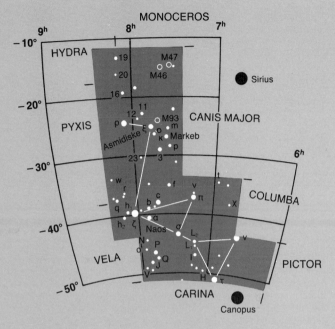

PYXIS

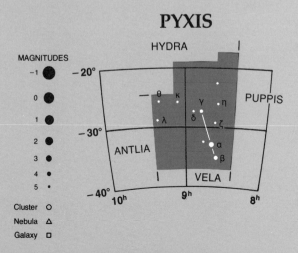

MAGNITUDES

-1
0
1
2
3
4
5

Cluster ○
Nebula △
Galaxy □

RETICULUM

(rē-tīk'-ū-lŭm)   Reticuli   Ret   *The Net*

This constellation of relatively faint stars is usually attributed to La-caille, but was defined earlier by one Isaak Habrecht of Strassburg, Germany. It is supposed to represent not a fishing net but a reticle, or grid, used to establish a scale in an eyepiece. This commemorates the work of Lacaille, who mapped previously uncharted stars in the southern hemisphere. Reticulum is just north of the Large Magellanic Cloud.

α is a G9III star, of magnitude 3.33, at a distance of 390 lt-yr from Earth.

β, a G9 star of magnitude 3.80, is at 76 lt-yr.

γ is a 4.46-magnitude star, spectral type M5, whose distance has not been determined.

δ is an M2 star, magnitude 4.41, also at an unknown distance.

ε, a sub-giant of spectral type K5, is of magnitude 4.42, at a distance of 80 lt-yr.

SAGITTA

(sa-jīt'-a)   Sagittae   Sge   *The Arrow*

This faint but distinctive constellation has been called Cupid's Arrow—the one he shot into the heart of Apollo, causing him to fall in love with the nymph Daphne. In another myth it is the arrow that Hercules used to kill the eagle of Zeus, and in still another, the arrow with which Apollo killed the Cyclops. Although none of its stars is bright, the pattern is easy to locate and does suggest an arrow. To find it, look just north of Altair.

α, although only of magnitude 4.37, is intrinsically very luminous, being of spectral type cF8. Its distance is 540 lt-yr.

β is a giant G7 star, magnitude 4.45, at a distance of 250 lt-yr.

γ is a giant M0 star, 192 lt-yr from Earth, of magnitude 3.71.

δ is a spectroscopic double star, with components of spectral types gM2 and A0. Their combined magnitude is 3.78. The system is at a distance of 408 lt-yr.

M71 is a 7th-magnitude globular cluster, at 18,000 lt-yr.

RETICULUM

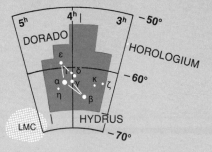

MAGNITUDES

SAGITTA

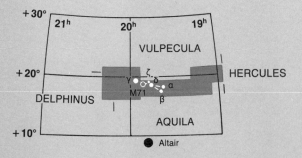

SAGITTARIUS

(săj'-ĭ-tā'-rĭ-ŭs)   Sagittarii   Sgr   *The Archer*

The "archer" dates from Mesopotamián mythology. The Greeks saw a centaur, usually Chiron, with bow and arrow pointed at Scorpius. Many call the central area "The Teapot." Sagittarius is in the zodiac, the Sun being here December 19 to January 19. Our galaxy's center is in this direction. The area is rich in star clouds and deep-sky objects.

α, Rukbat, the "archer's knee," is a B9 star, magnitude 4.11, at 250 lt-yr.

β, Arkab, "the tendon," is an optical double of types B8 and A9, magnitudes 4.24 and 4.51. β₁ is at 272, β₂ at 130, lt-yr.

λ, δ, and **ε** are Kaus Borealis, Kaus Media, and Kaus Australis, referring to the northern, middle, and southern parts of the centaur's bow.

σ is **Nunki** (an old Mesopotamián name), a B2 V star, magnitude 2.08, at 260 lt-yr.

M8 is the diffuse **Lagoon Nebula,** 40' across, at 5,100 lt-yr.

M17, the 7th-magnitude diffuse **Horseshoe Nebula,** is 20' across, 3,000 lt-yr distant.

M20 is the diffuse Trifid Nebula, 15' wide, 3,500 lt-yr from Earth.

M18, M21, M23, M24, and **M25** are 6th- and 7th-magnitude open clusters.

M22, M28, M54, M55, M69, M70, and **M75** are 5th- to 8th-magnitude globular clusters.

SCORPIUS

(skôr'-pĭ-ŭs)   Scorpii   Sco   *The Scorpion*

This is the scorpion that stung Orion to death. In classical times it was the largest constellation. Later, Libra was formed from its claws. It is part of the zodiac, the Sun being here November 23-30.

α, Antares, "rival of Mars," is a double, named from the bright red color of the primary. This M1 IB star is somewhat variable around magnitude 1.0. The secondary, 3" apart, is a faint green 5th-magnitude B4 star. The pair are 425 lt-yr distant. A 6-inch telescope should split them.

β, Acrab, is a triple. The primary and the secondary have a combined magnitude of 2.65; they are types B0 and B3, with 1" separation. The third star, 5th magnitude, is 13" away. The system is 650 lt-yr distant.

δ, Dschubba, "the forehead," is 590 lt-yr away, spectral type B0 V, magnitude 2.34.

υ, Lesath ("sting"), is a B2 IV star, magnitude 2.71, 540 lt-yr away.

λ, Shaula (also meaning "sting"), is a B1 V star, magnitude 1.62 at 325 lt-yr.

M6 is an open cluster, 6th magnitude, 1,500 lt-yr away.

M7 is a 5th-magnitude open cluster, 800 lt-yr from Earth.

M4 and **M80** are globular clusters of 6th to 7th magnitude.

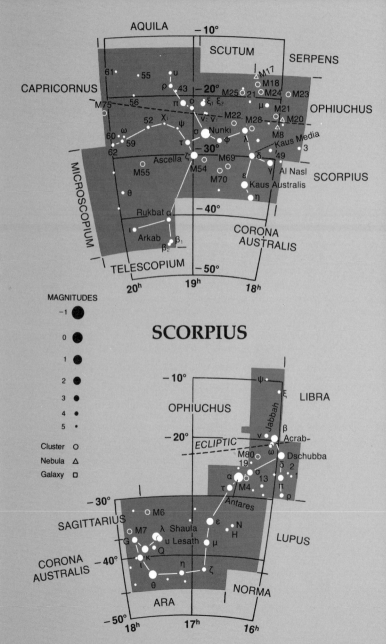

SAGITTARIUS

AQUILA
SCUTUM
SERPENS
CAPRICORNUS
OPHIUCHUS
MICROSCOPIUM
SCORPIUS
TELESCOPIUM
CORONA AUSTRALIS

−10°
−20°
−30°
−40°
−50°

20ʰ 19ʰ 18ʰ

MAGNITUDES
−1
0
1
2
3
4
5

Cluster ○
Nebula △
Galaxy □

SCORPIUS

OPHIUCHUS
LIBRA
ECLIPTIC
SAGITTARIUS
LUPUS
CORONA AUSTRALIS
NORMA
ARA

−10°
−20°
−30°
−40°
−50°

18ʰ 17ʰ 16ʰ

SCULPTOR

(skŭlp'-tēr)   Sculptoris   Scl   *The Sculptor*

This star group was originally *l' Atelier du Sculpteur*, "the Sculptor's Workshop," as named by Lacaille, but now is known simply as The Sculptor himself. It contains no very bright stars, but may be located by first finding Fomalhaut, then looking to the east. The south galactic pole is in this constellation at 0^h46^m, -27°, just above the star α. None of the stars is named.

α is a B5 star of magnitude 4.39 at 270 lt-yr.

β, a B9 star of magnitude 4.46, is 251 lt-yr distant.

γ is an sgG8 star, 155 lt-yr from us, of magnitude 4.51.

δ, at 163 lt-yr, is an A0 star of magnitude 4.64.

ζ is a B7 star of magnitude 4.99, at 988 lt-yr.

SCUTUM

(skū'-tŭm)   Scuti   Sct   *The Shield*

Hevelius named this small, faint group of stars Scutum Sobiescianum, "Sobieski's Shield." It was supposed to be the coat of arms of John Sobieski III (1624-1696), king of Poland. He defeated the Turks, who were marching on Vienna under the command of Kara Mustapha, on September 12, 1683. Later, Flamsteed shortened the name to Scutum. This group lies just north of Sagittarius and southwest of Aquila. It is conspicuous only because the Milky Way goes through this part of the sky. In addition to the bright M-objects, there are a number of fainter deep-sky objects visible in large telescopes.

α is a gK5 star of magnitude 4.06, at 204 lt-yr. Flamsteed called this star 1 in Aquila, before the 1930 codification of the constellations.

β is a star of magnitude 4.47 and spectral type cG7. It is 1,300 lt-yr from Earth, and was named 6 Aql in the catalog of Flamsteed.

γ is an A3 star of magnitude 4.73. It is 148 lt-yr from the solar system.

δ is a giant F4 star at 188 lt-yr. It varies around 5th magnitude. It is the prototype of a class of variable stars that change brightness by a few tenths of a magnitude over a period of a few hours. Such stars have about a tenth of the mass of the Sun. The light output varies as the star size changes.

M26 is a 9th-magnitude open cluster.

M11 is a very rich and famous 7th-magnitude open cluster 5,600 lt-yr away. The cluster is fan-shaped, 12'.5 wide (more than one-third the apparent size of the full Moon). It is an excellent object for binoculars or a small telescope.

SCULPTOR

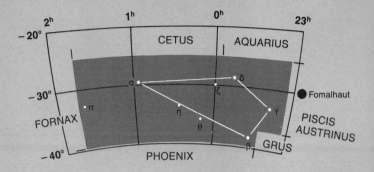

- 20°

2ʰ   1ʰ   0ʰ   23ʰ

CETUS   AQUARIUS

α   δ

- 30°   ζ   ● Fomalhaut

FORNAX   η   γ

π   θ   PISCIS
AUSTRINUS

- 40°   β   GRUS

PHOENIX

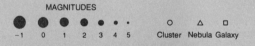

MAGNITUDES

-1   0   1   2   3   4   5   ○   △   □
Cluster   Nebula   Galaxy

SCUTUM

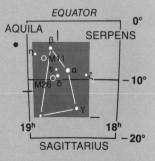

EQUATOR   0°

AQUILA   β   SERPENS

η   M11
ε   α   ζ
M26   δ   - 10°

γ

19ʰ   18ʰ   - 20°

SAGITTARIUS

SERPENS
(For this constellation see pp. 170-171)

SEXTANS
(sĕks'-tănz)   Sextantis   Sex   *The Sextant*

This little constellation was named by Hevelius to commemorate his large sextant, used by him in Danzig, 1658-1679, to chart the heavens. The sextant was destroyed in a fire. Sextans is found just south of Regulus.

α is an A0 star of magnitude 4.50, at 272 lt-yr.

β is a B5 star, 362 lt-yr off, with a magnitude of 4.95.

γ, an A0 star of magnitude 5.16, is 233 lt-yr away.

TAURUS
(tô'-rŭs)   Tauri   Tau   *The Bull*

In Greek myth it was this bull (Jupiter in disguise) that carried off Europa, beautiful daughter of the king of Phoenicia, to Crete. The figure was known also as the Cretan Bull and Egypt's Apis Bull. About 2000 B.C. the vernal equinox was here. Today the Sun is in Taurus from May 14 to June 21. The Bull's face is the Hyades cluster. On the Bull's shoulder are the Pleiades.

α is **Aldebaran**, "follower" of the Pleiades and eye of the Bull. It is between us and the Hyades, being only 68 lt-yr away. It is of spectral type K5III, magnitude variable about 0.85.

β is **El Nath**, "the butting one," the horn tip. It is of magnitude 1.65, spectral type B7III, 179 lt-yr away.

γ is a gG9 star of magnitude 3.86, at 142 lt-yr.

The Hyades, visible to unaided eyes, are "rainy stars," because their rise introduces autumn, the rainy season, or because they are the nymphs who wept over their brother Hyas, killed by a boar. The cluster is over 6° in diameter, 130 lt-yr away.

M45 is **The Pleiades** or **Seven Sisters,** daughters of Atlas, who holds the world on his shoulders. They were "sailing stars" for the Greeks. In Polynesian legend they were a bright star broken up by the god Tane for boasting. The group is 2° wide, 541 lt-yr away. Atlas and the mother, Pleione, are in the group. Only six are visible to unaided eyes; a small telescope shows about 100. The brightest Pleiads are: **Alcyone** (ăl-sī'-ŏ-nē), **Merope** (mĕr'-ŏ-pē), **Celano** (sī-lē'-nō), **Taygeta** (tā-ĭj'-ĭ-ta), **Sterope** (stĕr'-ŏ-pē), **Electra** (ĭ-lĕk'-tra), and **Maia** (mā'-a). Sterope is sometimes spelled Asterope.

M1 is the famous **Crab Nebula,** remnant of a brilliant supernova observed in China July 4, 1054. It was visible in daytime for months, at night for more than a year. A small telescope shows it, without detail. In it a rotating neutron star, or pulsar, has been detected—one of the fastest known. The nebula is about 5' wide, 4,000 lt-yr away.

SEXTANS

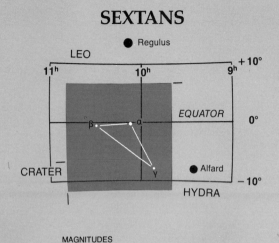

● Regulus

LEO

11ʰ   10ʰ   9ʰ   + 10°

EQUATOR   0°

β   α

CRATER   – 10°

γ

● Alfard

HYDRA

MAGNITUDES

●   ●   ●   ●   ●   ·   ·
−1   0   1   2   3   4   5

○ Cluster   △ Nebula   □ Galaxy

TAURUS

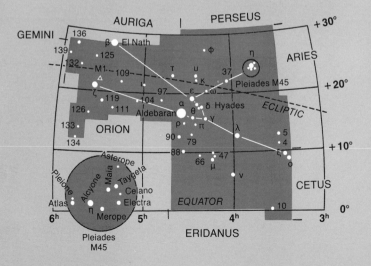

AURIGA   PERSEUS   + 30°

GEMINI   136   β El Nath   φ   ARIES

139   125   η

132   37   Pleiades M45

M1   109   τ   υ   κ   + 20°

△   ε   ω

ζ   119   104   97   *ECLIPTIC*

126   111   α   δ Hyades

Aldebaran   θ   γ

133   ρ   π   λ   5

134   ORION   90   79   4

88   47   ξ   o

66   μ

Asterope   CETUS

Pleione   Alcyone Maia Taygeta   ν   + 10°

Pleiades   Celano

Atlas   η   Electra   *EQUATOR*   10

Merope

Pleiades
M45   ERIDANUS

6ʰ   5ʰ   4ʰ   3ʰ   0°

ORION

TELESCOPIUM

(těl'-ē-skō'-pī-ŭm)　　Telescopii　　Tel　　*The Telescope*

This star group was originally the "Tubus Astronomicus," which La-caille formed between Sagittarius and Ara. There once was another constellation known as Telescopium Herschelii, to honor Sir William Herschel, but this group has passed into disuse. Telescopium contains no bright stars.

α is a star of magnitude 3.76, spec-tral type B6, at a distance of 652 lt-yr.

λ is a B9 star of magnitude 5.03. Its distance from Earth is unknown.

ε is a G5 star, magnitude 4.60, 297 lt-yr off.

ξ is an M2 star, magnitude 4.86, at a distance of 392 lt-yr.

ζ is a K0-type star of magnitude 4.14, at 148 lt-yr.

ι is a star of spectral type G9, mag-nitude 5.02, at a distance of 544 lt-yr.

TRIANGULUM

(trī-ăng'-gū-lŭm)　　Trianguli　　Tri　　*The Triangle*

Since any three stars in the sky can be connected to form a triangle, why has this particular group been seen as such for so long? Perhaps because it is almost isosceles. The ancients drew it as equilateral—and perhaps it was, then. The Greek astronomer Aratos thought it was the celestial representation of the island of Sicily, whose patron goddess is Ceres. In Bayer's time, some people called it the Christian Trinity or the Miter of St. Peter. Triangulum can be found just under Andromeda, above Aries.

α is a star of magnitude 3.42 and spectral type F6IV. It is at a dis-tance of 65 lt-yr.

β is a giant A5 star, magnitude 3.00, 140 lt-yr away.

γ is a A0 star, of magnitude 4.07, at a distance of 109 lt-yr from Earth.

M33 is a 6th-magnitude galaxy of type Sc, one of three spirals in the Local Group of galaxies, of which our Milky Way galaxy is one. The other spiral in the group is M31, the famous galaxy in Andromeda. M33 is 2.4 million lt-yr from us. It is about a degree wide, twice the apparent size of the full Moon.

TELESCOPIUM

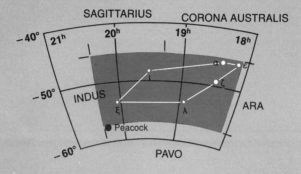

MAGNITUDES

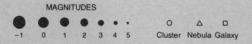

TRIANGULUM

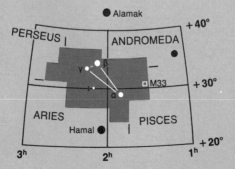

TRIANGULUM AUSTRALE

(trī-ăng'-gū-lŭm ôs-trā'-lē)   Trianguli Australis   TrA
The Southern Triangle

The depiction of this group is attributed to Pieter Theodore (16th century), but it was illustrated for the first time in maps by Bayer, in 1603. Its brighter stars are brighter than those of the northern Triangulum, recognized many centuries earlier by the ancients. To find the Southern Triangle, look for α and β Centauri, then go west-southwest through Circinus.

α is the brightest star of the constellation, magnitude 1.93, of spectral type K4III. It is 90 lt-yr from Earth.

β, an F0IV star of magnitude 2.84, is 42 lt-yr from our solar system.

γ, an A0V star, has a magnitude of 2.89 and is 113 lt-yr away. With α and β it makes up the main figure of the triangle.

δ is a G0 star of magnitude 4.03, at a distance of 130 lt-yr.

ε, a K0 star of magnitude 4.11, is almost exactly along the side of the triangle marked by the stars β and γ. It is 112 lt-yr away.

NGC 6205 is an open cluster with stars of 7th magnitude and fainter.

TUCANA

(tū-kā'-na)   Tucanae   Tuc   *The Toucan*

This constellation supposedly depicts the toucan, a well-known bird of South America. The name is from the language of the Tupi, an Indian tribe that European explorers found living along the coasts of Brazil and Paraguay and in the Amazon River Valley. The constellation looks little like its eponym, but is famous for the Small Magellanic Cloud, on its border with Hydrus, and for the globular cluster 47 Tucanae. Tucana's brightest star is only of 3rd magnitude.

α is a K4III star of magnitude 2.87. It is 62 lt-yr from Earth.

β is a triple star, with two components of magnitudes 4.52 and 4.48, types B9 and cA2, separation 27".1, at a distance of 148 lt-yr, plus a star of magnitude 5.16, spectral type A2, only 93 lt-yr away. The former pair are a system; the third star is merely a foreground object.

γ is an F0 star of magnitude 4.10, at a distance of 86 lt-yr.

δ is a B9 star, a double of combined magnitude 4.8, separation 6".8, 217 lt-yr distant.

ε is a B9 star of magnitude 4.71 at a distance of 233 lt-yr.

ζ is an F8 star of magnitude 4.34, 23 lt-yr away.

47 Tucanae (NGC 104) is one of the most famous globular clusters. At a distance of 16,000 lt-yr it appears to the eye as a fuzzy "star" of magnitude 4. In a small telescope it is a brilliant ball of stars about 44' wide.

TRIANGULUM AUSTRALE

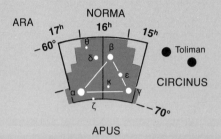

MAGNITUDES

-1 0 1 2 3 4 5

Cluster Nebula Galaxy

TUCANA

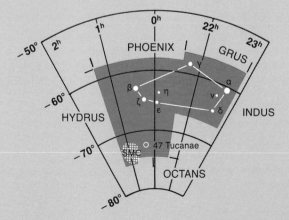

URSA MAJOR

(ûr'-sa mā'-jēr)   Ursae Majoris   UMa   *The Great Bear*

Most northern-hemisphere residents recognize the seven bright stars of the Big Dipper. The Pointers help us to find Polaris. Many diverse cultures have seen these stars as a bear, but the American Indians knew bears don't have long tails; so, for the Indians, the "tail" was three hunters chasing the bear, one carrying a pot (Alcor) to cook him in. In Europe, this star pattern is more commonly The Wagon or The Plow.

α, Dubhe ("the bear" in Arabic), is the Pointer nearer Polaris. It is a double (separation 1"), appearing as a K0III star of magnitude 1.79 at 104 lt-yr.

β, Merak ("loins of the bear"), is the second Pointer—an A1 V star of magnitude 2.37 at 78 lt-yr.

γ, Phecda ("thigh"), is an A0 V star of magnitude 2.44 at 90 lt-yr.

δ, Megrez ("root of the tail"), is an A3 V star of magnitude 3.30 at 63 lt-yr.

ε, Alioth (meaning uncertain), is an A0 V star of somewhat variable magnitude 1.78, 82 lt-yr distant.

ζ, Mizar (incorrectly, the "girdle"), is an A2 V star of magnitude 2.09

at 88 lt-yr. It has a companion, Alcor, 708" away. Often called "horse and rider," they test our eyesight.

η, Alkaid, or **Benetnash,** "end of the tail," is a B3 V star, magnitude 1.86 at 150 lt-yr.

M81 is an Sb galaxy, 7th magnitude, at 6.5 million lt-yr.

M82 is an irregular galaxy, 9th magnitude, at 6.5 million lt-yr.

M97, the 11th-magnitude planetary Owl Nebula, at 12,000 lt-yr, has 11th-magnitude central star.

M101 is an 8th-magnitude Sc galaxy at 14 million lt-yr.

URSA MINOR

(ûr'-sa mī'-nēr)   Ursae Minoris   UMi   *The Little Bear*

About 600 B.C. this group was suggested by Thales as a guide for Greek sailors. Some saw it as a dog's tail, revolving around Polaris, its tip. (From Greek *kynosouras*, "dog's tail," came Cynosure as a name for the pole star—and any other focus of interest.) Other cultures have seen this constellation as many kinds of objects, from jackal to jewels. As the Little Bear it is relatively modern.

α is **Polaris,** or **Stella Polaris,** the north pole star—a Cepheid variable, spectral type F8Ib, magnitude 1.99, at 782 lt-yr. A 9th-magnitude companion is 18" away.

β, Kochab, is a K4III giant, magnitude 2.07, 104 lt-yr off.

γ, Pherkad, is an A3II-III star, magnitude 3.04, distant 270 lt-yr.

δ, Yildun ("surpassing star"), is an A0 type, magnitude 4.44, at 233 lt-yr.

ε, a gG5 star, is an eclipsing binary of magnitude about 4, period 39.5 days, 300 lt-yr away.

ζ is an A2 star of magnitude 4.34, 217 lt-yr from us.

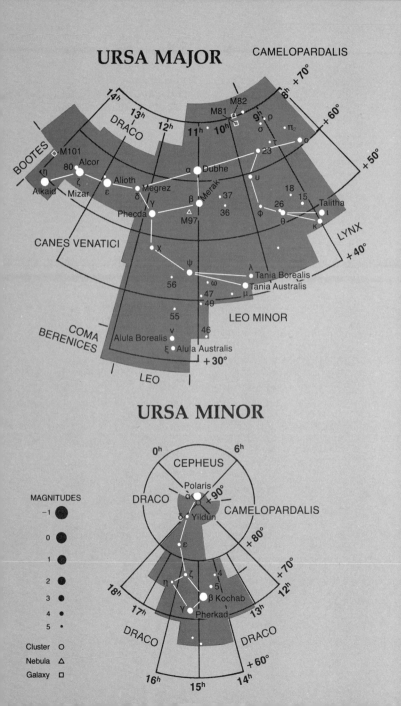

VELA

(vē'-la) Velorum Vel *The Sail*

This is the sail of *Argo Navis*. It is found above Carina, the Keel, and below Pyxis, the Compass. It contains several bright stars, but no α or β, since these designations went to other constellations when Argo was split up.

γ is **Alsuhail,** "brilliant" or "glorious"—a name sometimes given also to α Carinae. It is a very hot star, designated WC8. It is of magnitude 1.83 at 520 lt-yr.

δ is a double (separation 3") appearing as an A0V star, magnitude 1.95, 75 lt-yr off. Nearby is a companion, also a double.

κ is a B2IV star of magnitude 2.49, 470 lt-yr from us.

λ is a K4Ib star of magnitude 2.24, 750 lt-yr away.

φ, a B7 star of magnitude 3.70, at 1,800 lt-yr.

μ is a G5III star, magnitude 2.67, distant 108 lt-yr.

VIRGO

(vûr'-gō) Virginis Vir *The Maiden*

Recognized in ancient Egypt as a maiden, sometimes perhaps as the goddess Isis, and seen as a woman in many other early folklores, this constellation in classical myth was Astraea, goddess of justice, daughter of Jupiter and Themis. Next to her is Libra, the Scales. In other forms she was the daughter of Ceres, goddess of the harvest, holding a spike of wheat. Virgo is part of the zodiac, and the Sun is within this constellation from September 21 to November 1.

α is **Spica,** a "spike, or ear, of wheat." Worshipped by the ancient Egyptians, it is one of the brightest stars, an eclipsing variable of magnitude 0.91 to 1.01, spectral type B1V, at a distance of 260 lt-yr.

β, **Zavijava,** "angle" or "corner," is a dF8 star of magnitude 3.80, distant 32 lt-yr.

γ is **Porrima,** named for a minor goddess of justice. This is a double star (separation 3".9), each component being of type dF0 and magnitude 3.5, giving a combined magnitude of 2.76 to the system. It is at a distance of 32 lt-yr.

δ is a gM3 star of magnitude 3.66 at 181 lt-yr.

ε is **Vindemiatrix,** "grape gatherer," from the fact that just before vine-harvest time, this star rises in the morning. It is a G9II-III star of magnitude 2.83 at 90 lt-yr.

η is **Zaniah,** "kennel" (the Arabs saw this star as one of several forming a kennel). It is an A0 star of magnitude 4.00, at 142 lt-yr.

ι, **Syrma,** is "the train of the robe," a dF5 star of magnitude 4.16, 74 lt-yr away.

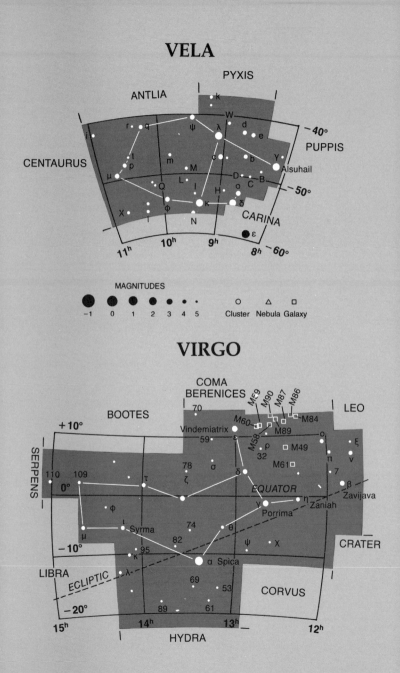

VOLANS

(vō'-lănz)   Volantis   Vol   *The Flying Fish*

This constellation, far to the south, was originally named Piscis Volans by Bayer, in 1603. It is between the bright star Miaplacidus (β Carinae) and the Large Magellanic Cloud. Its brightest stars are of magnitude 4. Having been figured in modern times, it has no mythology behind it, but immortalizes the flying fish seen by European explorers in tropical waters as they explored the New World.

α is an A5 star of magnitude 4.18, at a distance of 69 lt-yr from Earth.

β is a Kl star of magnitude 3.65 at a distance of 112 lt-yr.

γ is a double star. γ₁ is of magnitude 5.81 and spectral type G0. γ₂ is of magnitude 3.87 and spectral type F5. The separation is 13".7. The system is 130 lt-yr from our solar system.

δ is an F5 star of magnitude 4.02 at a distance of 1,090 lt-yr.

ε is a B8 star at a distance of 652 lt-yr. It has an apparent magnitude of 4.46.

ζ is a double (separation 6".1) appearing as a K0 star of magnitude 3.89. It is 117 lt-yr away.

VULPECULA

(vŭl-pek'-ū-la)   Vulpeculae   Vul   *The Fox*

Originally *Vulpecula cum Ansere*, the "Fox with the Goose," this constellation was named by Hevelius. It is between Cygnus, the Swan, on the north, and Delphinus and Sagitta on the south. Find the star Albireo (β Cygni), then look slightly to the southeast for Vulpecula. None of the stars is bright. Vulpecula is perhaps best known as the location of the Dumbbell Nebula.

α is a giant M1 star of magnitude 4.63, at a distance of 272 lt-yr. This is the only star in the constellation with a Greek-letter name.

13 Vulpeculae is of magnitude 4.50 and spectral type A0. It too is 272 lt-yr from Earth.

15 Vulpeculae is an A5 star of magnitude 4.74, at a distance of 130 lt-yr.

21 Vulpeculae· is an A3 star of magnitude 5.20, distant 272 lt-yr.

1 Vulpeculae is a B5 star of magnitude 4.60, at a distance of 326 lt-yr.

M27 is the famous **Dumbbell Nebula,** about 7' long, a planetary nebula of 8th magnitude, 3,500 lt-yr distant.

VOLANS

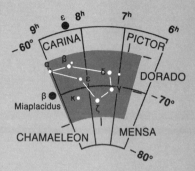

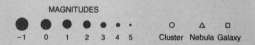
VULPECULA

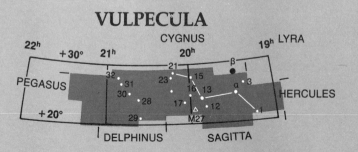

THE SOLAR SYSTEM

The solar system includes the nine major planets, at least five dozen satellites, hundreds of thousands of minor planets and meteoroids, and perhaps billions of comets, together with tenuous gas and dust pervading the entire space. On the opposite page, at top, is a depiction of the solar system with the planets revolving around the Sun. Because of distance and size relationships it cannot be to scale. The bottom illustration does show planet sizes compared to the Sun's. As shown, the plane of the solar system is tipped about 60° with respect to the plane of the Milky Way galaxy.

Planets can be classified as to physical properties. *Terrestrial planets* are relatively small, dense, rocky, with thin or nonexistent atmospheres. They include Mercury, Venus, Earth, Mars, and perhaps Pluto. *Gas-giant* (or *Jovian*) *planets* are large, of low density, with extensive atmospheres—Jupiter, Saturn, Uranus, and Neptune.

Planets are classified as to location in the solar system. The *inferior planets,* Mercury and Venus, are those with orbits smaller than the orbit of Earth. The *superior planets*—Mars, Jupiter, Saturn, Uranus, Neptune, Pluto—have orbits larger than Earth's.

All planets orbit the Sun counterclockwise as seen from the north side of the solar system (the side from which the north pole of Earth is visible). This is called *direct* motion. The minor planets and many satellites of planets have direct orbits, but some satellites and many comets have clockwise, or *retrograde*, orbits. For data on satellites, see page 257.

Some planets are not perfectly spheroidal; they are somewhat oblate, or flattened. Spherical planets have an oblateness of 0.

The *escape velocity* of a planet is the velocity necessary for an object at the planet's surface to escape from that planet without falling back. *Albedo*, or reflecting power, is the fraction of the incident sunlight the planet reflects. *Rotation* is the motion of a planet or other body around its own axis. *Revolution* is its motion in an orbit around another object. (For more on planetary motions, see pp. 246-248.) Planets are often compared with Earth. For reference, here are data on our planet:

> **Equatorial diameter:** 12,756 km.
> **Oblateness:** 0.0034. **Density:** 5.52.
> **Escape velocity:** 11.2 km/sec.
> **Inclination of equatorial plane to orbital plane:** 23.4°.
> **Eccentricity of orbit:** 0.017. **Albedo:** 37%
> **Average distance from Sun:** 149,597,870 km; 1 a.u.
> **Period of revolution:** Sidereal: 365.2422^d
> **Rotation period:** 24^h = 1 solar day
> **Atmosphere:** Fairly thin, mostly nitrogen and oxygen.
> **Satellites:** 1.

THE SUN

Always observe the Sun with care! See p. 206.

The Sun is our local star, a ball of gas radiating 4×10^{33} ergs per second. Earth receives about one two-billionths of this energy. The Sun is 865,000 miles across, about 109 times the size of Earth; it is about 93 million miles distant, and 107 similar suns could fit between it and the orbit of Earth. The Sun rotates counterclockwise (as seen from the north), the equatorial regions having a period of about 26 days, the polar regions about 37 days. We detect these relative motions by watching sunspots.

The source of the Sun's energy is *thermonuclear fusion:* the conversion of hydrogen into helium at the Sun's core, which releases energy. At the core the temperature is about 15 million degrees Kelvin, and the density of the gas is eight times that of gold. It takes millions of years for this energy to leak outward through the layers of gas surrounding the core. Theoretical models and studies of the solar system indicate the Sun has been shining for about 4½ billion years.

The *photosphere* is the "surface" of the Sun, where the gas becomes transparent enough to let light escape into space. A *sunspot* is a relatively cool region of the photosphere, produced probably by magnetic "storms." Whereas the photosphere is at about 6,000 degrees Kelvin, a sunspot may be a thousand or more degrees cooler. It appears dark, but would be bright if isolated from the contrasting surface. The number of sunspots varies in cycles averaging about 11 years. Peaks in numbers of sunspots have come as close together as 8 years and as far apart as 15. Counting and charting sunspots is a common activity of amateur astronomers.

The *chromosphere* is the layer immediately above the photosphere. It is called that because during eclipses, when the photosphere is hidden, the chromosphere appears red. The Sun's spectral lines are produced in this layer.

The *corona* is the Sun's outer atmosphere. Special telescopes called *coronagraphs* allow the inner part of the corona to be seen at any time. The outer part, stretching far out into the solar system, can be seen only during total solar eclipses. The corona is a tenuous gas with a temperature of millions of degrees.

Spurting above the photosphere are *prominences*—glowing jets of gas. Many are cloudlike, forming in the corona and "raining" back down onto the photosphere. They may be hundreds of thousands of miles long.

Active regions of the Sun sometimes produce *solar flares:* sudden bursts of visible and invisible light lasting seconds or minutes, best seen with special filters. They eject streams of charged particles which may, days later, strike Earth's higher atmosphere, causing aurorae or interruptions in radio communications.

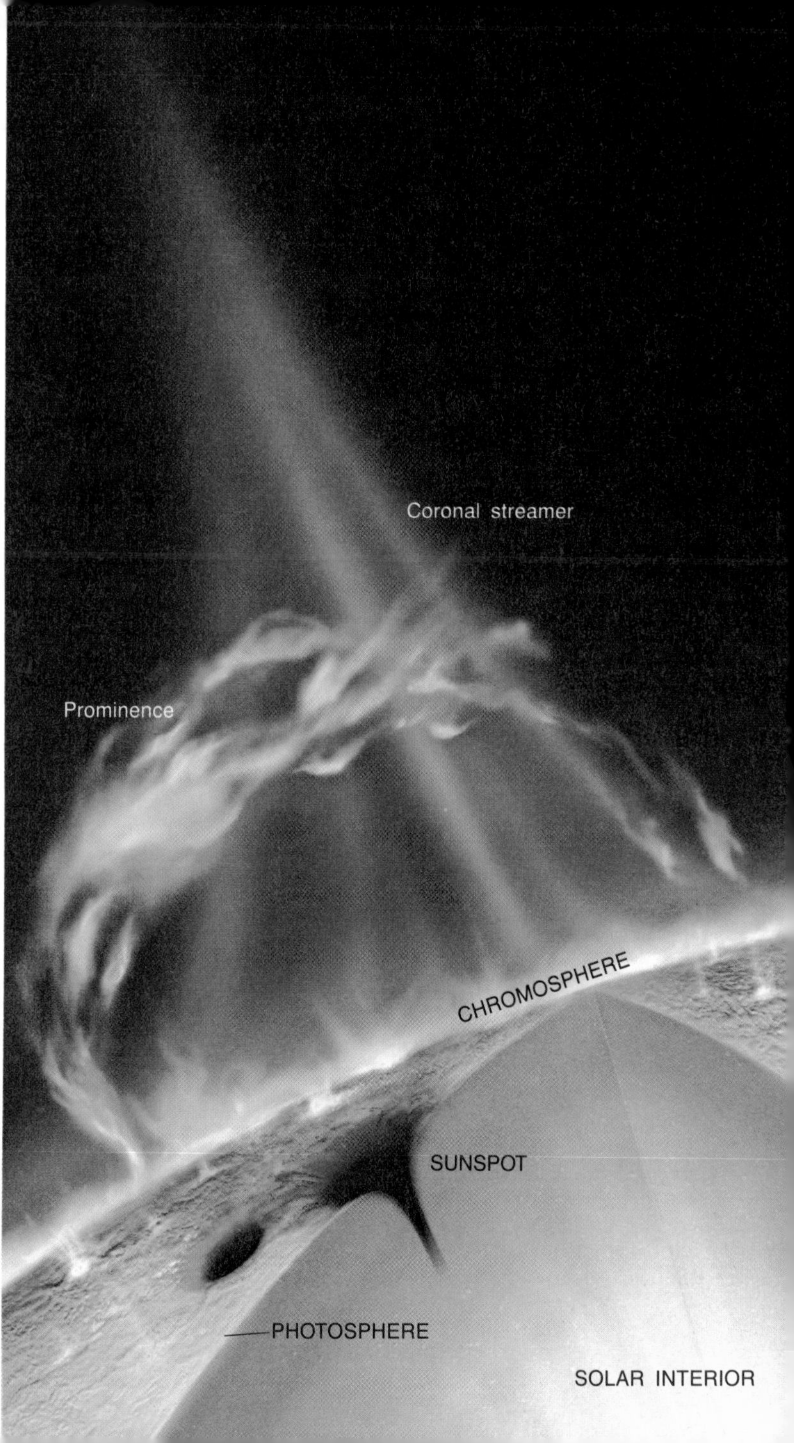

Coronal streamer

Prominence

CHROMOSPHERE

SUNSPOT

PHOTOSPHERE

SOLAR INTERIOR

SOLAR ECLIPSES

A solar eclipse occurs when the Moon blocks sunlight that normally falls on Earth. Besides being of some scientific interest, solar eclipses are perhaps Nature's most beautiful spectacle.

The Moon's shadow consists of two parts. The *umbra* is the dark inner portion, from the inside of which no part of the Sun's surface can be seen. The lighter, outer portion of the shadow is the *penumbra*. For a person within this, a part of the Sun's disk is visible.

A solar eclipse can occur only at the time of new moon, when the Moon is between Earth and Sun. Because the plane of the Moon's orbit is tilted slightly to the ecliptic plane, in most months the Sun-Moon-Earth alignment is only approximate, the Moon's shadow misses the planet, and no eclipse is visible from Earth. This situation is shown in the bottom illustration.

At the time of new moon a few times each year, the Moon's shadow does touch Earth. If only the penumbra touches Earth, people within the shadow will be able to see part of the Sun blocked off: a *penumbral* eclipse. Even when as much as 90% of the Sun's disk is covered, people may not notice the eclipse, because the rest of the Sun is so bright.

If the umbra passes across Earth, an umbral, or *total*, solar eclipse occurs. Only people within the umbra see totality; those outside see a penumbral eclipse or no eclipse at all. Both Moon and Earth are moving, so the tip of the umbra moves across the surface of Earth, tracing out the *eclipse path*. The shadow moves about 1,000 miles per hour eastward. The duration of totality is determined by the size of the shadow and the geometry of how it hits the surface of Earth. The maximum length is about 7½ minutes. During totality the Sun's disk is completely blocked and the beautiful corona is visible.

Since the Moon's orbit is elliptical, sometimes the umbra does not reach all the way to Earth. In such cases a thin ring, or *annulus*, of light is seen around the edge of the Moon at maximum eclipse. This is an *annular eclipse*.

As seen from Earth, the instant when the Moon's disk first touches the Sun's disk is called *1st contact*. The beginning of totality (if it occurs) is *2nd contact*. The end of totality is *3rd contact;* the end of the eclipse, *4th contact.* Just as the last part of the Sun's disk disappears, and again as it reappears, we may see the beautiful *diamond-ring effect*. Small spots of sunlight filtering through valleys at the edge of the Moon are known as *Bailey's beads*.

At least two solar eclipses occur each year, and a maximum of five are possible. As many as three eclipses may be total or annular, or none may be. Though solar eclipses are slightly more common than lunar eclipses, they are less familiar, because one must be inside the relatively narrow eclipse path to see the eclipse. On the average, you can expect the path of a solar eclipse to cross your location once every 360 years.

UMBRA

PENUMBRA

Location of totality
at this moment

TOTAL SOLAR ECLIPSE

Umbra too short
to reach Earth

ANNULAR
SOLAR ECLIPSE

Only penumbra
touches Earth

PARTIAL
SOLAR ECLIPSE

(Could also be early or late stage of annular or total eclipse)

NO ECLIPSE VISIBLE
FROM EARTH

Shadow misses Earth

HOW TO VIEW THE SUN SAFELY

Viewing the Sun directly without proper precautions can be very dangerous. The safest way, particularly for several viewers, is to make a *sunbox*—a sort of pinhole camera. Take a cardboard box, cut off one side, and tape a white piece of paper on the inside of one end. In the opposite end, cut a hole about 2 inches across and tape a piece of aluminum foil over the hole. Then very carefully punch a hole in the foil about ⅛ inch in diameter.

To use the sunbox, simply point it at the Sun so that the Sun's image falls on the paper. If the image is too faint, enlarge the hole. If the hole is too large, image quality will suffer. You might want to make a more durable opening by drilling the right-sized hole in a piece of aluminum or copper and then taping that over the cutout.

A more sophisticated projection technique is to use a simple lens, binoculars, or a telescope. DO NOT LOOK THROUGH IT. Use its shadow to get it pointed toward the Sun. Again use a white piece of paper as a screen. If using binoculars or a telescope with an eyepiece that has lenses cemented together, don't keep it pointed at the Sun long. You may crack the glass or fry the cement, and ruin the lens.

Special, usually expensive filters are available for advanced amateurs for direct viewing through a telescope. Use only filters made for this purpose; use them correctly. Don't improvise.

Neutral-density photographic filters are *not suitable* for visual viewing, either alone or with a telescope or through a camera with a telephoto lens. Such filters do not stop harmful infrared rays.

A filter made of film is safe for brief views of the sun. Take a roll of black-and-white (*not* color) film, expose it *completely*, and have it developed. Sandwich at least two layers of this film between glass, and tape the edges. To use it, first place the filter in front of your eye, then look at the Sun for no more than 5 seconds—sufficient to follow the progress of a solar eclipse.

For the few minutes of totality of a solar eclipse, when the Sun's disk is completely covered by the Moon, but *only* then, it is safe to view the Sun without filter. The shimmering corona sheds less light than the full Moon. Once the slimmest slice of Sun reappears, use the filters or sunbox again.

Some people have the idea that the Sun's rays are more dangerous during an eclipse. Not so; however, during an eclipse more people are looking at the Sun and so more eye injuries occur at these times. Take the proper precautions: the effects may be delayed, but you can become permanently blinded, and you may not even feel it happen. With the proper methods, though, there is no reason not to enjoy looking at the Sun.

PINHOLE
CAMERA
LUCIDA

PROJECTED IMAGE
OF SUN

Glass

Film

Glass

Solar filter
and aperture
stop

TABLE AND MAPS OF SOLAR ECLIPSES

The table below lists all solar eclipses for the period 1990-1999 and categorizes them as total (T), annular (A), or partial (P). A few unusual eclipses may begin as annular, become total for part of the path, and then become annular again (A-T). The portion of the eclipse that is total is not indicated in the maps.

The central lines of all total and annular eclipses through 1999 are shown with year, month, and day on the facing page. A triangle marks the beginning of each path; a bar marks the end. The regions of partial eclipse are not shown. The actual duration of the annular or total phase of the eclipse depends on where you are within the path.

More detailed information can be obtained from the references in the Bibliography. To do your own eclipse timing predictions, consult the *Canon of Solar Eclipses* listed there. Each astronomy magazine usually has a major article, sometimes several, on each upcoming total eclipse, giving detailed timings, weather predictions, and other information. Several companies offer tours and cruises to view total eclipses; read the advertisements in astronomy magazines.

SOLAR ECLIPSES, 1990-1999

Date	Type	Region of Visibility
1990 Jan 26	A	S South America, S Pacific
1990 Jul 22	T	NE Europe, NW North America, Hawaii
1991 Jan 15	A	Australia, New Zealand, S Pacific
1991 Jul 11	T	Hawaii, SW Canada, United States, Mexico
1991 Dec 21	P	Arctic, North and Central America, N Europe
1992 Jan 4	A	E Asia, Pacific, W North America
1992 Jun 30	T	South America, Africa, Indian Ocean
1992 Dec 23	P	NE China, Korea, Japan
1993 May 21	P	North America except SE, Arctic
1993 Nov 13	P	Antarctica
1994 May 10	A	NE Asia, North and Central America, Arctic
1994 Nov 3	T	Central and South America, S Africa
1995 Apr 29	A	Central and South America, Caribbean
1995 Oct 24	T	Arabia, Asia, Japan, Australia, Pacific
1996 Apr 17	P	New Zealand, S Pacific
1996 Oct 12	P	NE Canada, Greenland, Europe, N Africa
1997 Mar 8	T	E Asia, Japan, NW North America
1997 Sep 1	P	S Australia, New Zealand, S Pacific
1998 Feb 26	T	S and E North America, Central America
1998 Aug 21	A	S and SE Asia, Indonesia, Australasia
1999 Feb 16	A	Indian Ocean, Antarctica, Indonesia
1999 Aug 11	T	NE North America, Arctic, Europe, Arabia

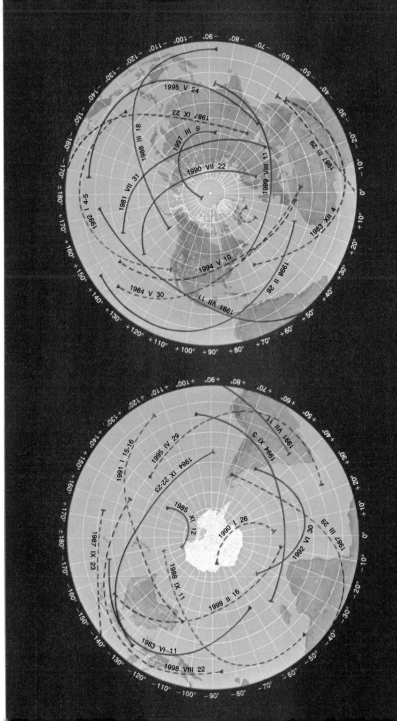

THE MOON

Our nearest celestial neighbor and companion is 2,160 miles in diameter, a fourth as large as Earth and the largest satellite compared to its planet in the solar system. The Moon orbits Earth at an average distance of 238,000 miles, and thus appears to be about ½° in diameter in the sky. The orbit is elliptical, so that the Earth-Moon distance varies by about 30,000 miles. The orbit is tilted about 5° to the plane of Earth's orbit.

After more than 4 billion years of tidal interaction with Earth, the Moon keeps one face permanently toward us. Thus the Moon rotates once on its axis each time it makes one revolution about Earth. Because of a slight wobble, called *libration*, what we can see adds up to about 59% of the total surface; the other 41% is never visible from our planet. As the Moon orbits, the entire half turned toward the Sun is illuminated, but the amount of this lighted half that we can see varies as the Moon revolves around Earth. Thus the Moon seems to change shape, or *phase*. The far side of the Moon is lighted part of the time and dark part of the time; hence it is incorrect to call it the "dark side."

The Moon takes 27d7h43m12s to make one complete orbit around Earth. This is called its *sidereal* period, for in this length of time it comes back to the same position relative to the background of distant stars. But in that time it will not be back to the same phase; its position relative to the Sun will have changed, because the Sun seems to move as Earth orbits it. The Moon takes about 2¼ days more than its sidereal period to return to the same phase. This longer period, referred to as the *synodic* period, is 29d12h44m3s in length.

New moon is the position, and time, when Earth, Sun, and Moon are roughly aligned. New moon is invisible, except during solar eclipses. *First-quarter moon* occurs when the visible side of the Moon is half illuminated, and occurs about a week after new moon. From new moon to *full moon*, when the entire visible face is bright, the Moon is said to be *waxing*. During the interval from either quarter moon to full moon, when the Moon is rounded on both sides, it is said to be in *gibbous phase*. About a week after full moon is *last-quarter phase*. The Moon is now *waning* toward new again.

Often when rising or setting the Moon appears larger than at times when it is higher in the sky. This apparent increase in size is an optical illusion perhaps due to our involuntary tendency to compare the size of the Moon with horizon objects.

Because the Moon approximately follows the ecliptic and thus the full Moon must be opposite the Sun, the full Moon in winter is high in the sky, and in summer low.

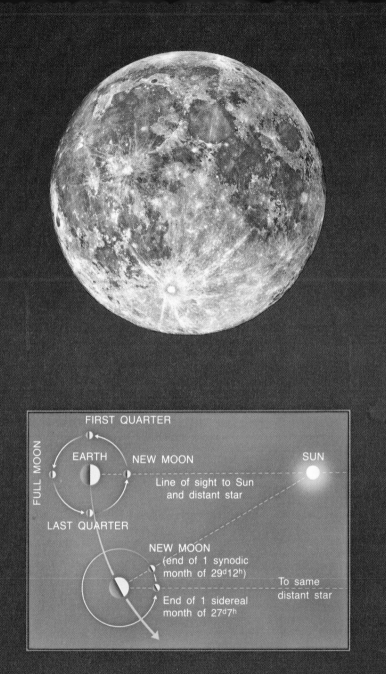

FIRST QUARTER

FULL MOON

EARTH

NEW MOON

SUN

Line of sight to Sun
and distant star

LAST QUARTER

NEW MOON
(end of 1 synodic
month of 29ᵈ12ʰ)

To same
distant star

End of 1 sidereal
month of 27ᵈ7ʰ

211

LUNAR FEATURES

Astronomers believe the Moon was fully formed three billion years ago, in the solar system's early days. Volcanic activity, meteoroid impacts, and fracturing of the crust created surface features which have changed little since they were formed because the Moon, having no atmosphere, lacks the erosional processes that would in time erase such features, as they have on Earth.

The Moon is close enough for detail to be studied even with binoculars; in a small telescope the Moon is very exciting. As the angle of sunlight falling on the Moon changes, aspects of lunar features also change. Through the centuries, people with a penchant for art have found the Moon a favorite subject for sketching.

For prolonged viewing of the Moon, if it is bright, use a neutral-density filter over the eyepiece, or reduce the light entering the telescope by placing a disk of cardboard with a hole in the center over the upper end of the tube. Then your vision will not be impaired.

Maria (plural of Latin *mare*, "sea") are not bodies of water, as thought by ancient astronomers, but broad plains—probably ancient lava flows. Darker than highland areas, maria are conspicuous, pocked with craters and dotted with mountains. Chains of maria cover much of the Moon's Earth-facing side. An example (opposite page) is Mare Serenitatis, the Sea of Serenity. Other "watery" terms for lunar features include Oceanus, Lacus (lake), Sinus (bay), and Palus (swamp). The English translations are used increasingly.

Craters, named mostly after scientists, range from very small to over 150 miles in diameter. Some were made by volcanoes, others by meteoroid impacts. Large circular features more than about 50 miles across are *walled plains.* Many craters have central peaks; walled plains do not. Craters usually have steeper sides and higher ringing mountains.

Mountains vary in size. Isolated peaks are common and with their high relief are interesting in small telescopes, especially when shadows are long. Many are named after Earthly mountains. *Valleys* are rare, especially large ones. Famous is the Alpine Valley, running through the lunar Alps on the edge of Mare Imbrium. *Rilles,* also called *clefts,* are elongated, winding, often deep features cutting across maria and crater floors. (Chains of craters may produce a similar effect.) *Rays,* light-colored linear areas radiating from young craters, are material thrown out by impacts of meteoroids.

Domes, difficult to see in small telescopes because they are low, are small rises on otherwise flat surfaces. Occasionally there is a small crater at the top of a dome.

Faults are fractures along which parts of the lunar surface are relatively displaced. A trough-like region sunk by faulting below the level of surrounding terrain is a *graben.* The most famous fault feature is the long cliff called the Straight Wall, near the crater Birt in Mare Nubium.

Mare Serenitatis, a lunar "sea"

The crater Copernicus

The Alps, a lunar highland region

The crater Tycho and surrounding highlands

213

FIRST-QUARTER LUNAR FEATURES FOR SMALL TELESCOPES

As the Moon waxes, features on this western, or first-quarter, side become visible. Each night, as the *terminator*—the line separating light and dark—moves over the features, their appearance changes because of the changing angle of illumination. It is along the terminator that features appear with maximum contrast.

The large "seas" make up the figure of a boy, the Jack of the Jack and Jill nursery rhyme: Mare Serenitatis is his head, Tranquilitatis his body, Nectaris and Foecunditatis his legs. Mare Crisium is the "pail" of the poem. As the Moon waxes, Jack "goes up the hill" before Jill, who is seen on the other side of the Moon.

MARIA AND OTHER LARGE FEATURES

A.	Mare Serenitatis	**H.**	Lacus Somniorum
B.	Mare Tranquilitatis	**J.**	Lacus Mortis
C.	Mare Nectaris	**K.**	Mare Undarum
D.	Mare Foecunditatis	**L.**	Mare Symthii
E.	Mare Crisium	**M.**	Mare Spumans
F.	Mare Frigoris	**N.**	Palus Somnii
G.	Mare Vaporum		

CRATERS AND WALLED PLAINS

1.	Hipparchus	**12.**	Bürg
2.	Horrocks	**13.**	Endymion
3.	Albategnius	**14.**	Julius Caesar
4.	Stöfler	**15.**	Aliacencis
5.	Maurolycus	**16.**	Werner
6.	Petavius	**17.**	Manilius
7.	Langrenus	**18.**	Delambre
8.	Eudoxus	**19.**	Macrobius
9.	Aristoteles	**20.**	Theophilus
10.	Hercules	**21.**	Posidonius
11.	Atlas	**22.**	Piccolomini

OTHER FEATURES

a.	Alpine Valley	**e.**	Alps Mountains
b.	Ariadaeus Rille	**f.**	Caucasus Mountains
c.	Pyrenees Mountains	**g.**	Appenines Mountains
d.	Taurus Mountains	**h.**	Rheita Valley

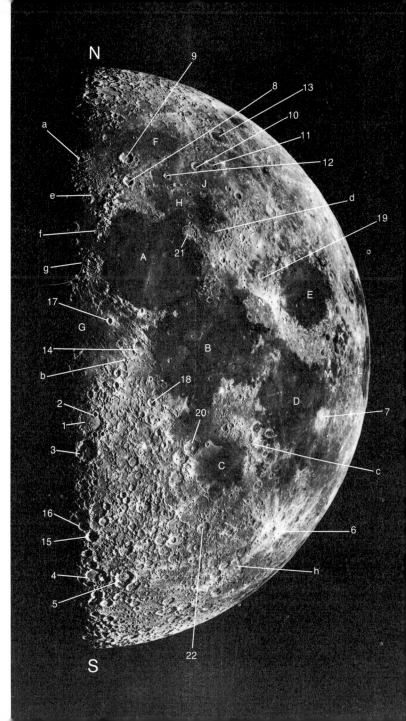

LAST-QUARTER LUNAR FEATURES FOR SMALL TELESCOPES

This eastern, or last-quarter, part of the Moon's visible side holds some of the most interesting features: the Alpine Valley; the Straight Wall; the crater Aristarchus, brightest on the Moon; the Straight Range; the peaks Piton and Pico, standing almost isolated on the plains of Mare Imbrium; and the huge walled plain Copernicus. The mountainous region in the Moon's southern hemisphere reveals different details each night as the angle of sunlight changes.

The last-quarter Moon shows Jill of the Jack and Jill nursery rhyme, though she is less well defined than Jack: Mare Imbrium is her head, Oceanus Procellarum her dress, and Mare Humorum her legs.

MARIA AND OTHER LARGE FEATURES

A. Mare Imbrium	**G.** Mare Nubium
B. Sinus Iridium	**H.** Sinus Aestuum
C. Mare Frigoris	**J.** Sinus Medii
D. Sinus Roris	**K.** Palus Nebularum
E. Oceanus Procellarum	**L.** Palus Putredinis
F. Mare Humorum	**M.** Mare Vaporum

CRATERS AND WALLED PLAINS

1. Clavius	**14.** Aristillus
2. Maginus	**15.** Epigenes
3. Tycho	**16.** Timocharis
4. Arzachel	**17.** J. Herschel
5. Alphonsus	**18.** Bianchini
6. Ptolemaeus	**19.** Copernicus
7. Alpetragius	**20.** Aristarchus
8. Albategnius	**21.** Herodotus
9. Herschel	**22.** Kepler
10. Pallas	**23.** Riccoli
11. Eratosthenes	**24.** Grimaldi
12. Archimedes	**25.** Gassendi
13. Autolycus	

OTHER FEATURES

a. Alpine Valley	**g.** Harbinger Mountains
b. Alps Mountains	**h.** Riphaeus Mountains
c. Piton Mountain	**i.** Straight Wall
d. Pico Mountain	**j.** Appenines Mountains
e. Straight Range	**k.** Spitzbergen
f. Jura Mountains	

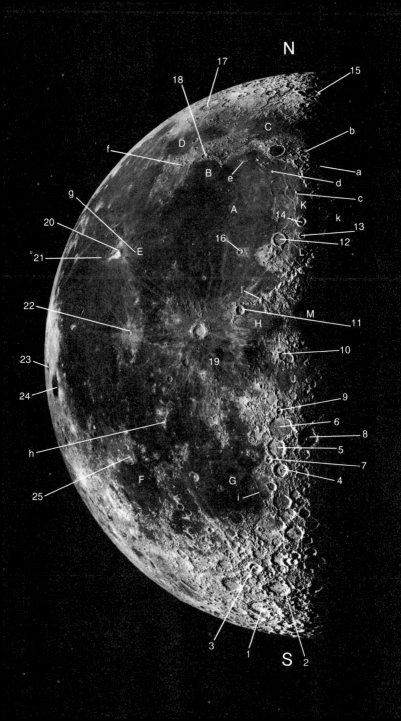

MOON PHASES

The Moon goes around Earth toward the east; it rises in the east and sets in the west; moonlight is just reflected sunlight. If you remember these three facts, you will easily be able to keep track of when the Moon is in the sky, its phase, and when it can be seen.

The Moon's orbit takes it around Earth so that its motion is toward increasing right ascensions. For a typical observer, looking up at the Moon in the southern sky, this motion is "to the left." It is easily visible over a period of a few hours, since the Moon moves, on the average, its own diameter each hour—that is, about ½° per hour.

The apparent motion of the celestial sphere carries the Moon 15° per hour in the opposite direction—westward. Thus the Moon rises in the eastern part of the sky and sets in the west, like most other celestial bodies.

At the time of new moon, the Moon will rise approximately at the same time as the Sun, and set approximately at the same time. During the daytime it will be above the horizon, though invisible. A day later, the Moon will have moved to the east of the Sun, so that it will follow the Sun across the sky, rising slightly after the Sun and setting slightly after it. Thus, the slim waxing crescent Moon can be seen low in the west just after sunset.

When a week has gone by, Sun, Earth, and Moon will be at right angles, and thus the Moon will be 90° eastward, or on the meridian, at sunset. The Moon will set about midnight. Each night it will rise (and set) later and later; on the average this retardation of moonrise amounts to 50 minutes per day, but varies considerably with the time of the year.

The waxing gibbous Moon is in the southeastern sky at sunset and sets after midnight. Two weeks after new moon, the full moon is opposite the Sun, rising at sunset and setting at sunrise. It is in the sky all night.

After full moon, the Moon rises after sunset; it is not visible in the early evening. Now it is waning. By the time of last-quarter phase, it rises at midnight and is on the meridian at sunrise, remaining in the sky (and visible, contrary to popular misconception) until about noon. The waning crescent rises later, gradually catching up with the Sun until another new moon, beginning another cycle.

The interval between one full moon and the next is a *synodic month*: 29d.53059. This is about 2¼ days longer than the time it takes for a full 360° revolution of the Moon around Earth. This revolution takes only 27d.32166: a *sidereal month*. The reason for the difference is that, as seen from Earth, the Sun also seems to move eastward in the sky against the background of stars. By the time the Moon has returned to the part of the sky where it was aligned with the Sun in the previous month, the Sun has moved on, and the Moon needs another 2¼ days to catch up.

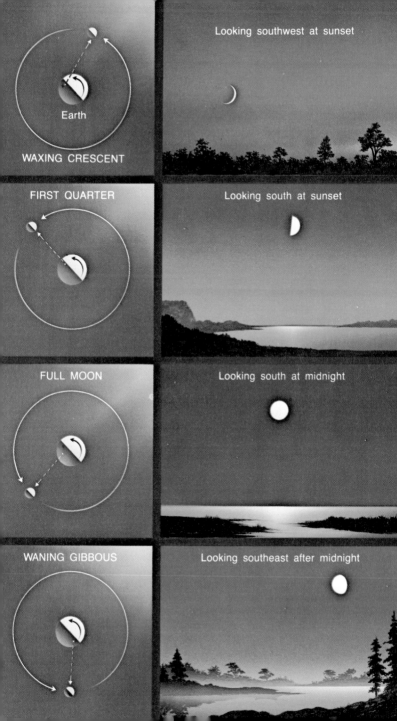

WAXING CRESCENT

Earth

Looking southwest at sunset

FIRST QUARTER

Looking south at sunset

FULL MOON

Looking south at midnight

WANING GIBBOUS

Looking southeast after midnight

TIDES

The basic theory of tides is fairly simple, but the interaction of seas with irregular shores and ocean bottoms makes actual tides complex. The arrows in the top diagram opposite show the gravitational attraction of the Moon on different parts of Earth. The longer the arrow, the greater the force. The Moon pulls more strongly on the near side of Earth than on the center, and more strongly on the center than on the far side. This "differential" causes the tides. Roughly speaking, water on the near side is pulled up and away from Earth, while Earth is pulled toward the Moon away from water on the far side.

At bottom left, solid arrows show the net force on the drop of water after the pull of the Moon on Earth's center has been subtracted. Dotted arrows are Earth's gravitational pull at the same locations. The solid arrows show how a drop of water will tend to move. Water tends to pull away from Earth on the sides of our planet closest to and farthest from the Moon; such locations have a high tide. Water flows away from regions in between, causing low tides. As Earth turns on its axis and the bulge of water produced by the Moon's pull stays roughly aligned between Earth and Moon, each location on Earth passes through two "high regions" and two "low regions" each day, with corresponding rises and falls of the water level. Because of the tilt of the Earth, a given location can experience two unequal highs and two unequal lows each day, as shown in the lower right illustration.

Because of the Moon's motion in its orbit around Earth, a lunar day is about 25 hours long rather than 24. Thus the time between successive high tides is about 12½ hours. Our diagram (lower right) shows why a place can have high or low tides of unequal height in one day.

The Sun is 27 million times more massive than the Moon but 400 times farther away. Since tidal forces decrease with the cube of the distance, the Sun's tidal effect is less than the Moon's.

When Sun and Moon are aligned at new moon and full moon, so that the total gravitational attraction is at maximum, tides are higher than normal—*spring tides*. Around the time of quarter or third-quarter moon, when Moon and Sun are at right angles, tides are lower than usual—*neap tides*. When the Moon is closest to Earth in its elliptical orbit, tides are higher, and when the Moon is more distant, lower. If full moon or new moon coincides with the Moon's perigee (closest approach to Earth), as it will a few times a year, tides are especially high.

All this is true for a spherical, water-covered planet. But Earth is only 70% covered, the oceans are of different depths, and water flows over bottoms variously shallow and deep, narrow and wide, steeply shelved or gradually sloping. Topographic features affect actual tides at any given location. For detailed local information, consult the *Tide Tables* published by the U.S. Hydrographic Office, or look in a local newspaper.

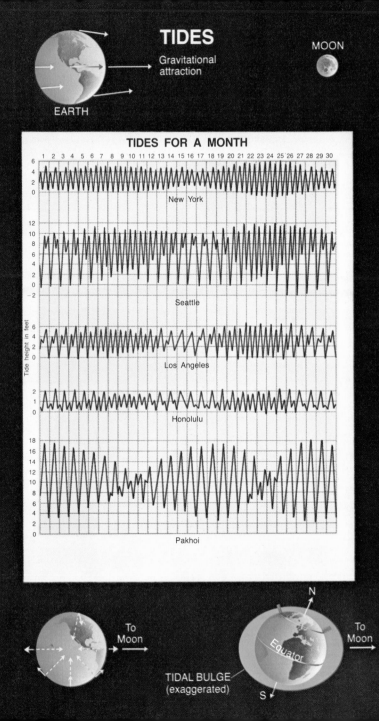

LUNAR OCCULTATIONS

As the Moon follows its orbit, it passes directly between us and certain objects farther away. If the object is a star or planet, this phenomenon is called an *occultation*. Thus a star can cast a shadow of the Moon on Earth's surface, and what you see from your location depends on where you are relative to that shadow (top illustration opposite). The bottom illustration shows four locations in such a shadow: Point 1 is at the *northern limit* of the occultation; Points 2 and 3 are at typical locations where a *total occultation* will occur; Point 4 is at the *southern limit*.

Occultations of bright stars are visible to the unaided eye, but are uncommon. One moment the star seems just beside the Moon; the next instant it is gone. How long it is out of sight depends on your location in the shadow. The maximum is about an hour, the time it takes for the Moon to move the distance of its own diameter against the background of stars. Stars, as points of light, disappear instantly; planets, being small disks of light, may take a few seconds.

Occultations are best watched through binoculars or a telescope. The middle illustration shows a telescopic view of how a star would appear to pass behind the Moon as seen from the four locations. It is easier to watch the events that occur at the dark edge of the Moon. Since the Moon moves eastward, the leading edge (at left for observers in the northern hemisphere) will be dark between new moon and full moon; the trailing edge will be dark between full moon and new moon. In a telescope, the star will appear to move relative to the Moon, but this is an illusion.

If you are viewing from a location just at the northern or southern limit, you may see a *grazing occultation* as the Moon just skims by the star. Within a narrow band, perhaps a mile wide, we can actually watch the star disappear and reappear repeatedly as it passes behind mountains and shines through valleys at the Moon's edge. The inset shows a greatly enlarged view of what happens. What we see is not the actual mountains and valleys, but the star blinking on and off. Often teams of amateur astronomers position themselves a few hundred meters apart along a line perpendicular to the northern or southern limit. If predictions are accurate, the observer at one end of the line will see a clean miss while others will see several disappearances and reappearances or a total, but brief, occultation.

Total occultations, in which the star disappears for some time, are useful for science. By timing these events carefully, relative to the locations of star, Moon, and observer, scientists increase their knowledge of lunar motions and distances between locations on Earth. Until the advent of spaceflight, grazing occultations were the only means of charting the peaks and valleys around the edges of the Moon. Magazines such as *Sky & Telescope* and the *Observer's Handbook* (see the Bibliography) list predicted times for many occultations; advanced amateurs can make their own predictions.

LUNAR OCCULTATIONS

Moon's shadow

To distant star

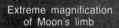

Extreme magnification
of Moon's limb

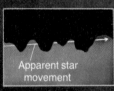

Apparent star
movement

1

2

3

4

LUNAR ECLIPSES

A lunar eclipse occurs when the Moon passes into Earth's shadow. Because the Moon's orbital plane is inclined to the ecliptic plane, an eclipse does not happen every month. The size of the shadow cast by Earth varies inversely as the distance between Earth and Sun, which changes during the year. Lunar eclipses occur only at full moon.

The dark inner portion of Earth's shadow, the *umbra*, is shaped like a cone, with its apex pointing away from the Sun. The tip is about 870,000 miles from Earth. At the Moon's distance from Earth, the diameter of the umbra is about 5,700 miles. The outer shadow, the *penumbra*, is a cone with the apex pointing toward the Sun and spreading out behind Earth. At the Moon's distance it is about 10,000 miles in diameter.

A *penumbral lunar eclipse* occurs when the Moon passes only through the penumbra. Such eclipses are hardly noticeable, for the diminution in light falling on the Moon is slight. A *partial lunar eclipse* occurs when the Moon passes through only part of the umbra. A *total eclipse* results when the Moon passes through the entire umbra. The first "touching" of the umbra by the Moon is *1st contact;* when just within the umbra, *2nd contact;* when just exiting the umbra, *3rd contact;* when just touching the umbra on the way out, *4th contact.* The maximum duration of a total eclipse, from 1st to 4th contact, is about 3 hours 40 minutes. A lunar eclipse is visible on Earth wherever the Moon is above the horizon. At least two, and a maximum of five, lunar eclipses can occur in a year.

As the Moon enters the umbra, from 1st to 2nd contact, it grows darker on its leading, eastward, edge, for as much as an hour. During totality, the only light from the Sun reaching the Moon is the small amount refracted through Earth's atmosphere around the edge of our planet, along the sunrise-sunset line. How much light reaches the Moon depends on the amount of cloud cover in those regions of Earth; sometimes the Moon is of a very bright coppery color; sometimes it is quite dark. Brightness cannot be predicted. Following 3rd contact, the leading edge brightens, and about an hour later the eclipse is over.

LUNAR ECLIPSES, 1990-1999

Date	E.S.T. of Mid-eclipse	Type	Duration of Totality	Date	E.S.T. of Mid-eclipse	Type	Duration of Totality
1990 Feb 9	1412	T	42^m	**1994** May 24	2232	P	—
Aug 6	0907	P	—				
1991 Dec 21	0534	P	—	**1995** Apr 15	0719	P	—
1992 Jun 14	2358	P	—	**1996** Apr 3	1911	T	86^m
Dec 9	1845	T	74^m	Sep 26	2155	T	70^m
1993 Jun 4	0802	T	96^m	**1997** Mar 23	2341	P	—
Nov 29	0126	T	46^m	Sep 6	1347	T	62^m
				1999 Jul 28	0634	P	—

P = partial eclipse; T = total eclipse

224

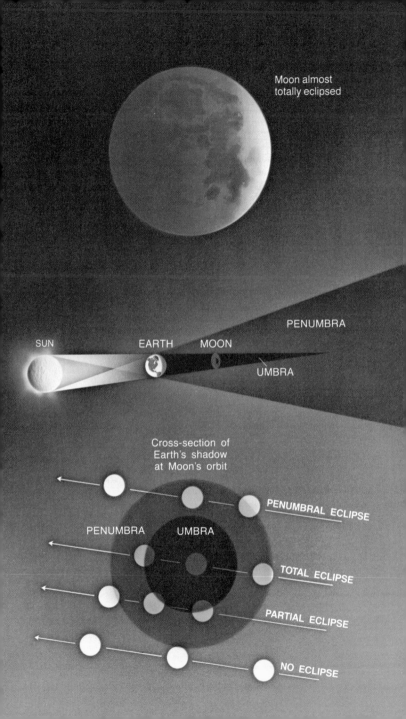

Moon almost
totally eclipsed

PENUMBRA

SUN   EARTH   MOON

UMBRA

Cross-section of
Earth's shadow
at Moon's orbit

PENUMBRAL ECLIPSE

PENUMBRA   UMBRA

TOTAL ECLIPSE

PARTIAL ECLIPSE

NO ECLIPSE

METEORS, METEOROIDS, AND METEORITES

A *meteor* is a flash of light in the sky produced by a fragment of rock or metal from space as it hurtles into our atmosphere, at a speed of about 25 miles per second, and is heated to incandescence by friction. Such fragments while in space are called *meteoroids*. One that has not disintegrated or vaporized during its plunge, but has landed on the ground is a *meteorite*. On the average clear, dark night, 5 to 10 meteors per hour can be seen. At times of meteor showers there are many more (see p. 228).

The best time to see meteors is after midnight. Then Earth is catching up to some of them and we are on the "leading edge" of our planet. In the top diagram opposite, the blue line is the midnight line. Past midnight, as Earth rotates we are "looking out the front windshield" and see more meteors, just as when driving in rain we see more drops hitting the windshield than hitting the rear window. For clarity the meteors are shown here as streaks of light in outer space; in reality they occur close to Earth's surface, mostly about 100 miles up.

Most visible meteors are caused by meteoroids about the size of a grain of sand. Brighter ones may be the size of a pea. Occasionally larger ones enter the atmosphere and become very bright; these are called *fireballs*. Fireballs that explode with an audible pop or bang are called *bolides*. Even faint meteors may leave trails lasting seconds or minutes. The middle drawing here is of a fireball seen by the artist over New Jersey.

Meteoroids amounting to over 400 tons of material land daily on Earth. Much is in the form of micrometeoroids, too small to make a visible flash, simply drifting down to the ground. Meteorites land all over Earth, but the chances that you will ever see one on the ground and recognize it are very, very small. If you find an object you suspect is a meteorite, take it to a local planetarium, observatory, or university geology department for examination. Very few specimens turn out to be real meteorites.

Most meteorites that fall are *stony*. A few are a mixture of nickel and iron; they are called *irons*. A very few may be *stony-irons*. Most commonly found are irons, because they resist weathering better than stones and are more conspicuous. They are heavy and may be pitted like slag from a furnace. Rarely, a meteor has been seen and the meteorite from it picked up, and in all such cases that are known the meteorite has been stony.

Earth is pockmarked with craters, or astroblemes, from large meteorite impacts in the past. Most famous in the United States is the Meteor Crater, near Flagstaff, Arizona, about ¾ mile across. Some older craters, tens or even hundreds of miles across, are difficult to recognize because of their size and because they have been nearly obliterated by ages of erosion.

226

METEORS

FIREBALL

Iron meteorite

Stony meteorite

Stony-iron meteorite
cross section

METEOR SHOWERS

A lone, random meteor seen in the night sky is called *sporadic.* Such meteors are produced by meteoroids arriving at Earth from all directions. Many meteoroids are known to be orbiting the Sun in streams which follow the orbits of comets. If such a comet orbit crosses the orbit of Earth, whenever Earth passes that point in its own orbit, once (or sometimes twice) a year, the number of meteors seen will increase. Thus occurs a *meteor shower,* during which all the meteors seem to come from one direction in the sky, radiating outward from a region called the *radiant.*

In the lower diagram, the view from Earth's surface is of a series of meteors that seem to streak radially from near a single point in the sky. (In reality there is no plane as depicted, and meteors actually are observed much closer to the surface than shown.) Meteors are seen all over the sky, not just in the direction of the radiant, and their paths may be long or short, but if the paths are traced backward, they intersect at the radiant.

A meteor shower is named for the part of the sky in which the radiant lies. Thus there are Perseids from Perseus, Lyrids from Lyra. Each shower has a peak date, perhaps varying a day or so from year to year, on which the number of shower meteors is at maximum. For a few days before and after the peak date, the number of shower meteors is greater than the number of sporadic meteors, though less than the peak number. Variations in the peak number occur from year to year depending on how the meteoroids are distributed in the stream.

The following table lists some major annual meteor showers. The duration is the total number of days, centered on the peak date, within which the number of shower meteors is one-quarter or more of the peak number.

MAJOR METEOR SHOWERS

Peak Date	Name (radiant)	Visible per Hour (est.)	Duration (days)
January 4	Quadrantids	40	2.2
April 21	Lyrids	15	4
May 4	Eta Aquarids	20	6
July 28	Delta Aquarids	20	14
August 12	Perseids	50	9.2
October 21	Orionids	25	4
November 3	South Taurids	15	?
November 16	Leonids	15	?
December 13	Geminids	50	5.2
December 22	Ursids	15	4

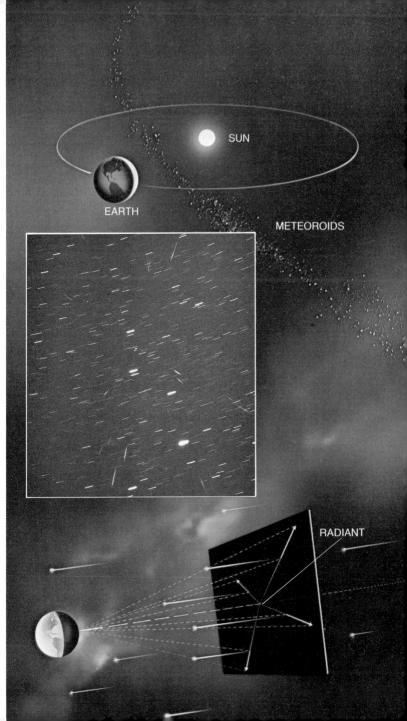

SUN

EARTH

METEOROIDS

RADIANT

MERCURY

Our solar system's innermost planet was noticed by the ancients and named for Roman mythology's messenger of the gods, because of its relatively swift motion across the sky. The symbol for this planet (☿) is a stylized caduceus, the staff entwined with serpents, often carried by the god Mercury and later a symbol of the healing arts.

Equatorial diameter: 4,878 km (0.382 × Earth's).
Oblateness: 0. **Mass:** 0.055 × Earth's. **Density:** 5.43.
Surface gravity: 0.38 × Earth's. **Escape velocity:** 4.3 km/sec.
Inclination of equatorial plane to orbital plane: near 0°.
Inclination of orbital plane to ecliptic plane: 7.0°.
Eccentricity of orbit: 0.206.
Av. distance from Sun: 57.9 million km; 0.387 a.u.
Period of revolution: Synodic, 116^d. Sidereal, 87^d.97 (0^y.241).
Rotation period: 58^d.46 (0^y.161). **Albedo:** 11%.
Atmosphere: Extremely thin, mostly hydrogen and helium.
Satellites: None.

Mercury appears as a fairly bright point of light visible for only about a half-hour, just after sunset or before sunrise, during a few periods each year. It is thus always close to the horizon and never more than about 18° from the Sun in the sky. Because its orbit is between Earth's orbit and the Sun, it shows phases, which repeat with a synodic period of 116 days. A 3- or 4-inch telescope may show phases. Mercury takes 44 days to go from greatest eastern elongation (seen as an "evening star"—see p. 248) to greatest western elongation (seen as a "morning star"), and another 72 days back to greatest eastern elongation. Some elongations are more favorable for viewing than others, because of the planet's orbital inclination to the ecliptic.

Once in a decade or so, at the time of inferior conjunction (the time when the planet is between Earth and Sun) Mercury can be seen to cross the face of the Sun. The crossing is called a *transit*. Because of the orbital inclination, a transit does not happen every month. Observers with proper equipment for viewing the Sun (CAUTION—see p. 206!) can then see Mercury as a small dot, like a sunspot but slowly moving across the solar disk. The next transits will be on November 6, 1993, and November 14, 1999.

Earth-based telescopes can detect no detail on Mercury, but space-craft photographs show a surface pitted with craters, somewhat like the Moon's. For many years it was thought that Mercury keeps one face permanently toward the Sun; in 1965, however, it was found that this is not so. Mercury has no atmosphere. Its magnetic field is very weak.

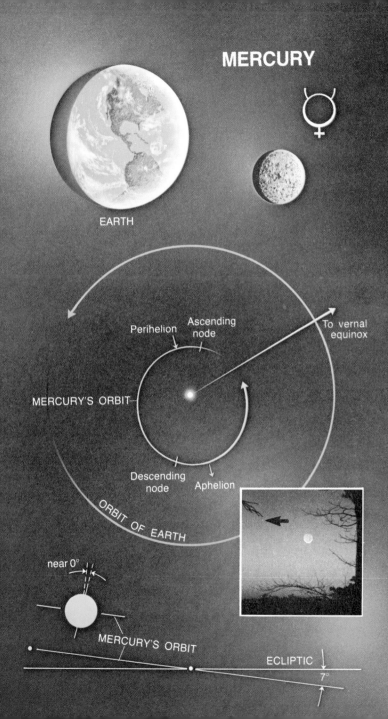

MERCURY

EARTH

Perihelion • Ascending node

To vernal equinox

MERCURY'S ORBIT

Descending node

Aphelion

ORBIT OF EARTH

near 0°

MERCURY'S ORBIT

ECLIPTIC

7°

VENUS

Venus is at times the brightest planet in the sky—brighter than all celestial objects except the Sun and Moon. Named for the goddess of love and beauty, it is the planet that comes closest to Earth. Its symbol (♀) derives from a stylized hand mirror, and is the universal symbol for female.

> **Equatorial diameter:** 12,104 km (0.948 × Earth's).
> **Oblateness:** 0. **Mass:** 0.8150 × Earth's. **Density:** 5.24.
> **Surface gravity:** 0.91 × Earth's. **Escape velocity:** 10.4 km/sec.
> **Inclination of equatorial plane to orbital plane:** 177.3°
> **Inclination of orbital plane to ecliptic plane:** 3.4°.
> **Eccentricity of orbit:** 0.007.
> **Av. distance from Sun:** 108.2 million kilometers; 0.723 a.u.
> **Period of revolution:** Synodic, 584^d. Sidereal, 224^d.70 (0^y.615).
> **Rotation period:** 243^d.017, retrograde. **Albedo:** 65%.
> **Atmosphere:** Very thick, mostly carbon dioxide.
> **Satellites:** None.

Venus is the most conspicuous morning and evening star, reaching a magnitude of -4 at brightest. Like Mercury, it has an orbit within Earth's orbit around the Sun and so shows phases. Reaching a maximum elongation of 48°, it is much more easily visible than Mercury. Venus takes about 144 days to go from greatest eastern elongation ("evening star") to greatest western elongation ("morning star"), and then about 440 days to return to greatest eastern elongation. Its greatest brightness occurs about 36 days before and after inferior conjunction, when the planet is in a crescent phase.

In a small telescope, Venus appears larger than Mercury, because it is larger and comes closer. Its apparent size varies from about 11″, when at superior conjunction, to about 1′, at inferior conjunction. This planet rarely transits the Sun's disk. The next transits will be June 8, 2004, and June 7, 2012; then not again until 2117 and 2125.

The beauty of Venus is only skin-deep. The clouds in its atmosphere, which are responsible for its high reflecting power and thus its brightness, prevent any glimpse of the surface in visible light. Maps made by radar reveal a stony surface with large valleys, plateaus, and craters. Atmospheric pressure at the surface is about 90 times that of Earth's atmosphere, and the surface temperature is about 900°F. The atmosphere is mostly carbon dioxide, with clouds largely of sulfuric and hydrochloric acid droplets! Large-scale weather patterns are visible only in ultraviolet light, some sweeping around the planet at more than 700 kilometers per hour. The planet has no appreciable magnetic field.

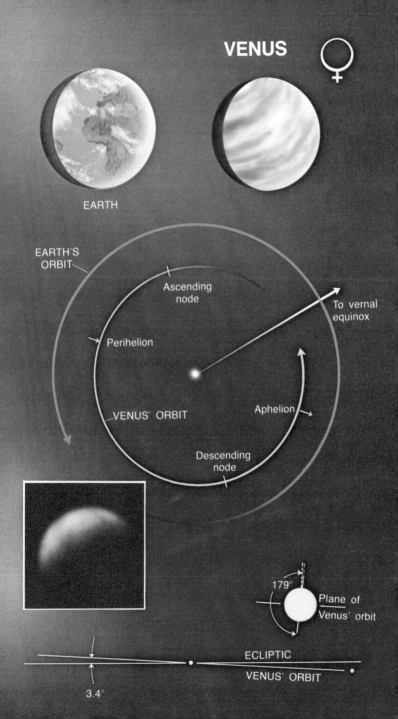

VENUS

EARTH

EARTH'S
ORBIT

Ascending
node

To vernal
equinox

Perihelion

VENUS' ORBIT

Aphelion

Descending
node

179°
Plane of
Venus' orbit

ECLIPTIC

VENUS' ORBIT

3.4°

MARS

The reddish color of this planet reminded the ancients of blood, and they named it for the god of war. The Greeks called it Ares. To this planet, more than any other, people have looked for life beyond Earth. Mars' symbol (δ) is a stylized spear and shield, the symbol for male.

Equatorial diameter: 6,787 km (0.533 × Earth's).

Oblateness: 0.0052. **Mass:** 0.1074 × Earth's. **Density:** 3.94.

Surface gravity: 0.38 × Earth's. **Escape velocity:** 5.0 km/sec.

Inclination of equatorial plane to orbital plane: 25.2.

Inclination of orbital plane to ecliptic plane: 1.8°.

Eccentricity of orbit: 0.093.

Av. distance from Sun: 227.9 million kilometers; 1.524 a.u.

Period of revolution: Synodic, 780^d. Sidereal, 686.90^d (1^y.881).

Rotation period: 24^{h}37^{m}23^s (1^d.0260) **Albedo:** 15%.

Atmosphere: Thin, mostly carbon dioxide.

Satellites: 2. (See p. 257)

The orbit of Mars around the Sun is outside the Earth's orbit, and so Mars can appear in the sky at elongations up to 180°. Its eccentric orbit brings it to within 56 million kilometers of Earth every 17 years, but there is an opposition (Sun-Earth-Mars alignment) every 780 days. At its brightest, Mars is distinctly red. Its magnitude at oppositions ranges from −1 to −2.8 depending on the Earth-Mars distance at those times.

The atmosphere is about 0.005 as dense as Earth's, being mostly carbon dioxide, with little oxygen or water vapor. The polar caps, which wax and wane with the Martian seasons, are partly frozen water, partly dry ice. Enormous dust storms, with high winds, sweep the planet. A few thin clouds are seen occasionally. Spacecraft photographs show a surface with high mountains, old craters, and deep valleys. Ancient volcanoes produced a few enormous peaks, such as Olympus Mons, largest known volcanic mountain in the solar system. No canals or other artificial features exist. Photographs show channels that seem to have been made by flowing water, but no surface water can exist now under Mars' low atmospheric pressure. Long ago Mars may have had a thicker atmosphere, allowing open water to exist on the surface. A small telescope shows little more than the polar caps and vague lines and patches representing the varied terrain.

The two tiny satellites are probably captured asteroids. The inner one, Phobos, orbits the planet so fast a person on Mars would see it rise in the west and set in the east, despite the planet's eastward rotation. Deimos has a period of revolution not much more than the planet's rotation period and thus would rise only once every 5½ days.

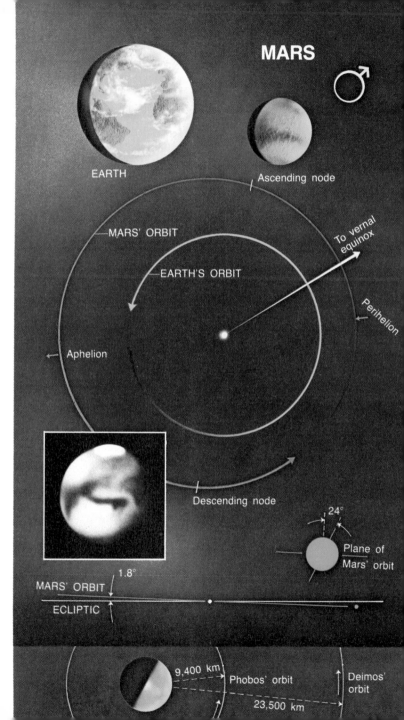

MARS

♂

EARTH

Ascending node

MARS' ORBIT

EARTH'S ORBIT

To vernal equinox

Perihelion

Aphelion

Descending node

24°

Plane of Mars' orbit

1.8°

MARS' ORBIT

ECLIPTIC

9,400 km   Phobos' orbit   Deimos' orbit

23,500 km

JUPITER

The largest of the planets is named for the king of the Roman gods, the Greek Zeus. Jupiter rules over a miniature "solar system" of satellites and is the prototype of the "Jovian," or gas-giant, planets. Its symbol (♃) is a stylized lightning bolt.

Equatorial diameter: 142,800 km (11.23 × Earth's).

Oblateness: 0.0588. **Mass:** 317.8 × Earth's. **Density:** 1.33.

Surface gravity: 2.87 × Earth's. **Escape velocity:** 59.6 km/sec.

Inclination of equatorial plane to orbital plane: 3.1°.

Inclination of orbital plane to ecliptic plane: 1.3°.

Eccentricity of orbit: 0.048. **Albedo:** 52%.

Av. distance from Sun: 778.3 million km; 5.203 a.u.

Period of revolution: Synodic, 399^d. Sidereal, 11^y.86.

Rotation period: 9^{h}55^{m}30^s (0^d.4041) (varies with latitude).

Atmosphere: extensive; mostly hydrogen, helium; some methane, ammonia.

Satellites: 4 large, 12 small. (See p. 257)

Because of its large size and high albedo, Jupiter can at times reach a magnitude of about −2.6. It is almost as bright as Venus, which is always much closer to Earth. Being a superior planet, Jupiter can appear in opposition to the Sun, a good position for viewing. A small telescope will show the "surface"—actually, the cloud tops of bright belts and dark zones parallel to the equator. Seeing the Great Red Spot, apparently a storm in Jupiter's atmosphere, is a challenge.

Jupiter may have no distinct solid surface. If one exists, it is very deep below the thick atmosphere. Pressures rise to millions of atmospheres, and temperatures to tens of thousands of degrees, near the center; hence the planet must have an extremely dense core.

Jupiter has an extremely strong magnetic field, and huge, powerful radiation belts. The planet emits radio waves that can be detected on Earth. Lightning has been observed in Jupiter's atmosphere.

In a small telescope, the four large "Galilean" satellites (those discovered by Galileo) can be seen to change their location with respect to the planet in a few hours and from night to night. The 12 smaller satellites and the faint ring discovered by the Voyager spacecraft are not visible in small telescopes.

The four large satellites are of planet size. Io, about the size of Pluto, shows extensive volcanism, not of liquid rock but of liquid sulfur shooting many kilometers into space. Europa, possibly largely water with an icy crust, is larger than Pluto and slightly smaller than the Moon. Ganymede, larger than Mercury, is three-fourths the size of Mars, and the second largest satellite in the solar system. Callisto, also probably mostly water, is about the same size as Mercury. None of the other satellites is larger than 200 kilometers in diameter, and it is likely that the outer ones are captured asteroids.

JUPITER

EARTH

Perihelion

Ascending node

To vernal equinox

JUPITER'S ORBIT

EARTH'S ORBIT

Descending node

3.1°

Plane of Jupiter's orbit

JUPITER'S ORBIT — 1.3°

ECLIPTIC

Jupiter V II IV   XIII X   XII   VIII

I III   Average distance of satellites   VI VII   from planet, to scale   XI IX

Amalthea
V

Io
I

Europa
II

Ganymede
III

Callisto
IV

Leda
XIII

Himalia
VI

Lysithea
X

Elara
VII

Ananke
XII

Carme
XI

Pasiphae
VIII

Sinope
IX

Satellite sizes compared to our Moon

Moon

SATURN

Saturn, the Greek Chronos, was the Roman god of time and agriculture. The planet marked the edge of the known solar system until the discovery of Uranus, late in the 18th century. As we see Saturn, it is the slowest-moving of the naked-eye planets. Its symbol ($\hbar$) is a stylized scythe.

> **Equatorial diameter:** 120,000 km (9.41 × Earth's).
>
> **Oblateness:** 0.11. **Mass:** 95.16 × Earth's. **Density:** 0.70 (it would float on water!).
>
> **Surface gravity:** 1.08 × Earth's. **Escape velocity:** 35.6 km/sec.
>
> **Inclination of equatorial plane to orbital plane:** 26.7°.
>
> **Inclination of orbital plane to ecliptic plane:** 2.5°.
>
> **Eccentricity of orbit:** 0.056. **Albedo:** 47%.
>
> **Av. distance from Sun:** 1,427 million km; 9.539 a.u.
>
> **Period of revolution:** Synodic, 378^d. Sidereal, 29^y.46.
>
> **Rotation period:** 10^{h}39^m (0^d.444) (varies with latitude).
>
> **Atmosphere:** Extensive, mostly hydrogen and helium.
>
> **Satellites (17):** 5 larger than 1,000 km; 12 others. (See p. 257.)

Saturn reaches only 1st magnitude at brightest, when at opposition, every 12½ months. The only satellites brighter than 10th magnitude are Titan and Rhea.

Saturn's rings are swarms of small particles orbiting the equator of the planet in a flat plane. Three major rings and many fainter ones exist, only the former being visible in small telescopes. Some of the rings are "braided"; others have a dynamical interaction with some of the smaller satellites. Every 15 years, Earth passes through the plane of the rings, so that the rings, being edge-on to our line of sight, become invisible in small telescopes for a short time.

Saturn's atmosphere is extensive, but its low density implies that the planet contains little solid material. Since the planet is farther from the Sun than Jupiter is, and thus subject to less variable heating, its atmosphere is less active than Jupiter's.

Of Saturn's 17 satellites, only 5 are larger than 1,000 kilometers: Tethys, Dione, Rhea, Titan, and Iapetus. Titan, the largest satellite in the solar system at 5,550 kilometers across, is larger than Mercury.

Titan is also the only satellite to have a thick atmosphere, mostly of nitrogen, with small amounts of methane and other gases. It is about 200 kilometers thick, and at the surface has a pressure about that of atmospheric pressure on Earth. We can observe faint clouds floating in the atmosphere. Sunlight converts some of the methane into hydrocarbon particles of smog. The smog is so dense it obscures the surface from our view. We are not sure if Titan's surface is solid or liquid.

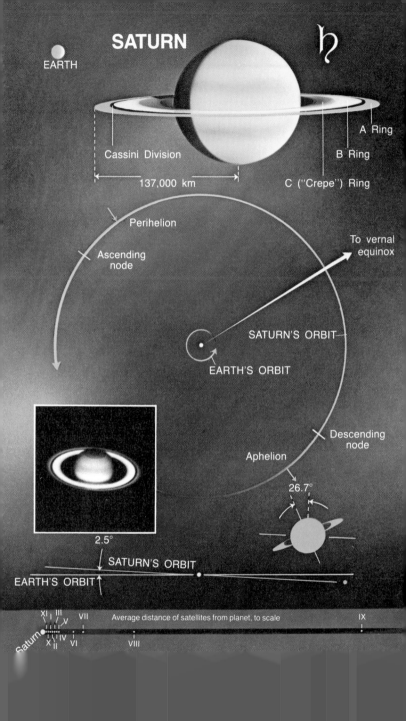

SATURN

EARTH

A Ring

Cassini Division

B Ring

137,000 km

C ("Crepe") Ring

Perihelion

Ascending
node

To vernal
equinox

SATURN'S ORBIT

EARTH'S ORBIT

Descending
node

Aphelion

26.7°

2.5°

SATURN'S ORBIT

EARTH'S ORBIT

Average distance of satellites from planet, to scale

Saturn

XI III   VII   IX

X   II   IV   VI   VIII

V

URANUS, NEPTUNE, PLUTO

These outer planets were named for a trio of Roman gods: Uranus, the god of the heavens (hence, of astronomy); Neptune, god of the ocean depths; and Pluto, god of the underworld. Uranus' symbol is a combination of the sun sign and a spear (⚨); Neptune's symbol (♆) is a trident; Pluto's (♇) is a combination of the first two letters of its name as well as the initials of the noted astronomer Percival Lowell, who started the search for this planet. Uranus and Neptune are gas giants; Pluto, with a density nearly equal to that of water, may be largely ice.

Uranus was discovered by chance in 1781 by Sir William Herschel (1738-1822). When it did not move as the laws of celestial mechanics predicted, astronomers calculated that a more distant planet could be perturbing it. A search turned up Neptune in 1846. Another, similar search led to the discovery of Pluto in 1930.

All three are relatively unrewarding for amateur observers. Uranus, with its magnitude of about 6, is barely visible to the unaided eye under the best conditions, and in a small telescope is but a tiny greenish disk. Neptune, at about 9th magnitude, requires binoculars; it too is a tiny greenish disk in a telescope. Pluto, with a magnitude of only 15 or so, requires a telescope of at least 16 inches to be seen even as a point of light. The table below summarizes data known so far for these planets.

THE OUTER PLANETS

	Uranus	Neptune	Pluto
Equatorial diameter	51,200 km (3.98⊕)*	48,500 km (3.81⊕)	3,000?
Oblateness	1/45	1/40	?
Mass	14.5⊕	17.2⊕	0.0026⊕?
Density	1.30	1.76	1.1?
Surface gravity	0.91⊕	1.19⊕	0.03⊕?
Escape velocity	21.3 km/sec	23.8 km/sec	?
Rotation period	$16^h48^m(0.7^d)$	16^h3^m	$6^d09^h18^m$
Inclination of equator to orbital plane	97.9°	29.6°	?
Albedo	51%	41%	50%?
Av. distance from Sun	19.182 a.u. 2,869,600,000 km	30.058 a.u. 4,496,600,000 km	39.785 a.u. 5,965,200,000 km
Sidereal period	$84^y.01$	$164^y.79$	247^y
Eccentricity of orbit	0.047	0.009	0.254
Inclination of orbit to ecliptic plane	0.8°	1.8°	17.1°
Synodic period	370^d	367^d	367^d
Known satellites	15	8	1

*The symbol ⊕ is for Earth; 3.98⊕ means "3.98 times Earth's."

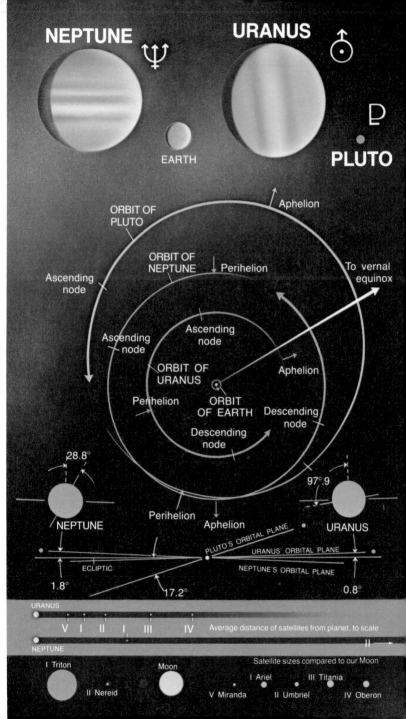

NEPTUNE Ψ

URANUS ⊙↑

EARTH

♇ PLUTO

ORBIT OF PLUTO

Aphelion

ORBIT OF NEPTUNE

Perihelion

To vernal equinox

Ascending node

Ascending node

Ascending node

Aphelion

ORBIT OF URANUS

ORBIT OF EARTH

Perihelion

Descending node

Descending node

Perihelion

Aphelion

28.8°

NEPTUNE

97°.9

URANUS

1.8°

17.2°

0.8°

PLUTO'S ORBITAL PLANE

URANUS' ORBITAL PLANE

NEPTUNE'S ORBITAL PLANE

ECLIPTIC

URANUS

V I II I III IV   Average distance of satellites from planet, to scale

NEPTUNE

II

Satellite sizes compared to our Moon

I Triton

II Nereid

Moon

I Ariel

III Titania

V Miranda   II Umbriel   IV Oberon

ASTEROIDS

Asteroids are *minor planets* (the preferred term—an older term is *plane-toid*). The largest asteroids are Ceres, 1,000 kilometers in diameter; Pallas, 590 km; Juno, 300; Vesta, 550; and Bamberga, possibly larger than Pallas. The sizes range downward to grains of sand. Most of the minor planets have orbits that lie between those of Mars and Jupiter, in what is called the *asteroid belt*. About 2,000 minor planets have orbits known well enough to be cataloged, and they are given numbers in order of discovery. Optionally the discoverer may propose a name. Thus, there are minor planets named for mythological persons, plants, pets, girl friends, and other assorted eponyms.

Some of the asteroids have orbits which carry them within the orbits of the inner planets. If the orbit crosses that of Earth, the asteroid is called an Apollo-object, after the first one discovered. Some even go within the orbit of Mercury. Two groups of minor planets, at the *Lagrangian points* in Jupiter's orbit—that is, the points 60° ahead of and behind the giant planet—are called the *Trojan asteroids*.

The top illustration shows some of the orbits of representative minor planets in plane view. The bottom illustration shows the inclinations of their orbits to the ecliptic. The fact that some minor planets have highly eccentric orbits and high inclinations leads some astronomers to think that not all minor planets are the result of collisions between larger objects, or leftover material from the formation of the solar system, but are rather the remains of comets.

Few minor planets ever reach naked-eye brightness. The larger ones when relatively near Earth can be spotted in binoculars or a small telescope if the observer has star charts with coordinates and knows when and where to look. In a telescope's field of view, over a period of a few hours an asteroid may be seen to move. On time-exposure photographs these bodies appear as short streaks, whereas the stellar images are points.

The largest of the minor planets are more or less spherical, since with their large mass they have sufficient gravitational attraction to be drawn into that shape. The smaller objects are more likely to be irregular. None of the asteroids can hold an atmosphere. In general, asteroids are completely solid bodies somewhat like meteoroids, many of which are "chips" resulting from asteroid collisions. Some asteroids are rocky, consisting mostly of silicates; some metallic, mostly iron and nickel; and some combine rock and metals. Possibly in the future, within a few decades, some minor planets can be moved to the vicinity of Earth to be mined for raw materials.

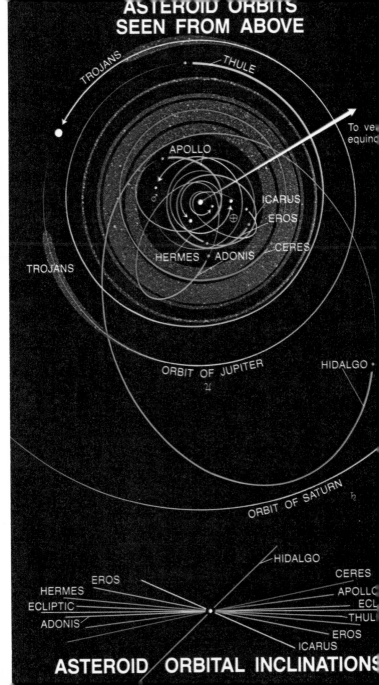

ASTEROID ORBITS
SEEN FROM ABOVE

TROJANS

THULE

APOLLO

To ve
equino

ICARUS

EROS

CERES

HERMES • ADONIS

TROJANS

ORBIT OF JUPITER
♃

HIDALGO •

ORBIT OF SATURN ♄

HIDALGO

EROS

CERES

HERMES

APOLLO

ECLIPTIC

ECL

ADONIS

THUL

EROS

ICARUS

ASTEROID ORBITAL INCLINATIONS

COMETS

A comet may appear anywhere in the sky at any time, looking like a fuzzy "star," perhaps with a tail. Many comets, racing in from far reaches of the solar system, go around the Sun, then head outward again. Comets probably are remnants of the systems's original material.

The *head* of a comet, perhaps a few kilometers wide, is probably a ball of ices—water, ammonia, methane, carbon dioxide, and other substances—mixed with rocks and dust. As a comet approaches the Sun, generally nearer than the orbit of Mars, some of the ice sublimates, freeing gases and solids which re-emit or reflect sunlight to form a hazy glow, the *coma*, around the head. Solar radiation pushes gases and dust away from the head, making a *tail* or tails that stream behind as the comet nears the Sun, and precede as it races away (illustration, lower left). Some comets have little or no tail, but lengths of 30 to 50 million miles are common. How long a tail looks is influenced by our angle of view (illustration, lower right). With each passage around the Sun, a comet loses material, and thus may survive only a few hundred passes.

Only comets that contain much gas and dust and that pass relatively near the Sun develop large tails. "Short-period" comets remain within the inner part of the solar system; for example, Encke's, with its period of 3.3 years—the shortest known. Others have periods of up to millions of years, and orbits far beyond Pluto's. Comets have all degrees of orbital inclination (illustration, center), and some orbits are retrograde. A periodic comet may look different each time it returns.

Astronomers discover, or rediscover, a dozen or so comets each year. Most are faint. Each is initially designated by its year and order of discovery; for example, 1981a, the first discovered in 1981. Later it is given a permanent designation with a roman numeral indicating its order of perihelion passage (passage nearest the Sun), for example, 1910 II. A comet may be named for as many as three discoverers.

About once each decade an unpredictable bright comet appears. The only predictable bright one is Halley's, which was very bright in 1910, but less so at its latest appearance in 1986.

Almanacs and astronomical handbooks list periodic comets that are likely to be visible, with probable dates of their return.

KEY TO COMET ORBITS

1. Encke 1977 XI **2.** Tuttle 1967 V **3.** Helfenzrieden 1766 II **4.** Pons-Winnecke 1976 XIV **5.** Barnard-1 1884 II **6.** Holmes 1892 III **7.** D'Arrest 1976 XI **8.** Brooks-2 1889 V **9.** De Vico 1846 VIII **10.** Lexell 1770 I **11.** Finlay 1974 X **12.** Denning-Fujikawa 1881 V **13.** Fay 1977 IV **14.** Barnard 3 1892 V **15.** Spitaler 1980 VII **16.** Pigott 1782 **17.** Tempel 1871 II **18.** Blanpain 1819 IV **19.** Biela 1852 III **20.** Brooks-1 1886 IV **21.** Wolf 1976 II **22.** Tempel-1 1977 I **23.** Brorsen 1979 I **24.** Tempel-2 1977 d **25.** Halley 1910 II

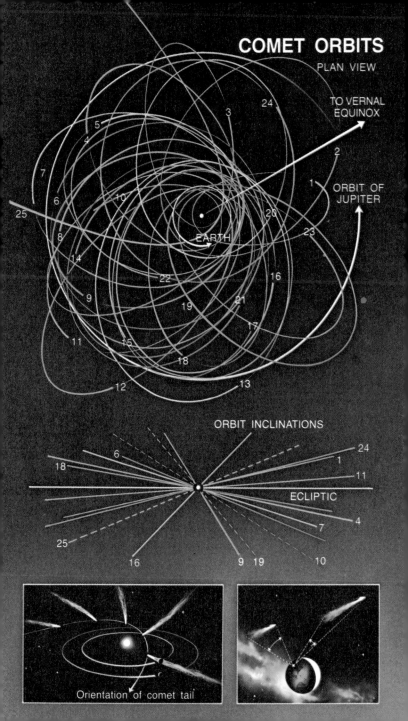

COMET ORBITS

PLAN VIEW

TO VERNAL EQUINOX

ORBIT OF JUPITER

EARTH

ORBIT INCLINATIONS

ECLIPTIC

Orientation of comet tail

ORBITS

The path of one body moving around another is called an *orbit*. For bodies, such as planets, subject to the law of gravity, the shape of an orbit is one of the conic sections, so called because they can be produced by the intersection of a cone and a plane. *Open*, or *unbounded*, orbits are those of objects not permanently bound to one another by gravitation; each orbit has at least escape velocity with respect to the other. *Closed* orbits are those which are periodic — that is, traversed periodically.

The German astronomer Johannes Kepler (1571-1630) deduced the rules of closed orbits, called *Kepler's laws*:

I. *The orbit of a planet above the Sun lies in a plane and has the shape of an ellipse with the Sun at one focus of the ellipse. The other focus is empty.*

An ellipse can be drawn by placing a loop of string over two tacks spaced some distance apart; these points are the *foci*. Keeping the string taut, move a pencil around the loop and it will draw an ellipse. The size of the ellipse is denoted by the *semimajor axis*, which is half the length of the longest dimension. The shape is measured by the *eccentricity*, usually denoted by e, which is a number between 0 and 1. When the eccentricity is 0, the figure is a circle, corresponding to the case in which the two tacks are in the same place.

II. *A simple relation exists between the size of the orbit and the period of revolution. If A, the semimajor axis, is measured in astronomical units and the period, or time, T, in years, for the planets the relation is $T^2 = A^3$.* (To remember which is squared and which is cubed, recall the area of Broadway and 42nd Street in New York: Times Square!)

III. *A line from the Sun to the planet sweeps across equal areas of space in equal intervals of time.* This means that a planet moves faster as its orbit brings it closer to the Sun (nearest point is the *perihelion*) and slower as it moves farther away (farthest point is the *aphelion*).

These laws are exactly true only for the case of just two orbiting bodies, but are a good approximation for the real solar system because the Sun is so much more massive than the planets, and the planets are relatively far apart, so that their perturbations due to their mutual gravitational attraction are small.

Several more variables are needed if you want to predict the position of a planet in the sky. One is *inclination*, or angle between the plane of the orbit and the ecliptic: it is called i. Another, denoted by the symbol Ω, is called the *longitude of the ascending node*: it gives the celestial longitude of the point where the plane of the orbit crosses the ecliptic. The *longitude of the perihelion*, denoted ω, is the angle in the orbital plane from the ascending node to the perihelion direction. The actual angle from perihelion to where the planet is at any given time is called the *true anomaly, f*. The other angles are defined in the lower illustration. For more details, consult one of the references in the Bibliography.

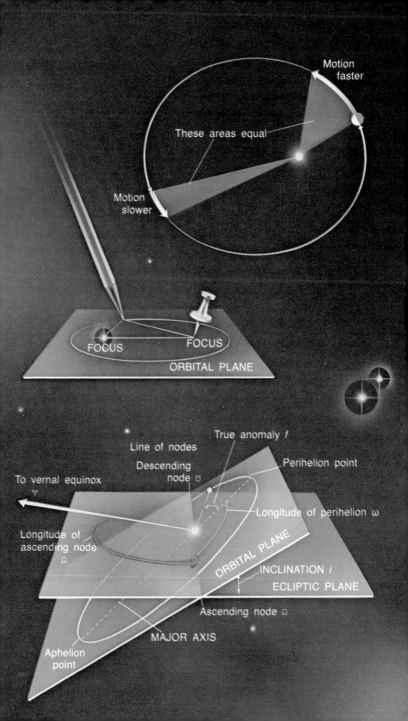

Motion
faster

These areas equal

Motion
slower

FOCUS FOCUS

ORBITAL PLANE

True anomaly *f*

Line of nodes

Descending
node ☊

Perihelion point

To vernal equinox
♈

Longitude of perihelion ω

Longitude of
ascending node
☊

ORBITAL PLANE

INCLINATION *i*

ECLIPTIC PLANE

Ascending node ☊

Aphelion
point

MAJOR AXIS

PLANETARY PHENOMENA

Positions of planets in their orbits are usually described relative to Earth, since that is our viewpoint. Planets with orbits within Earth's orbit around the Sun are called *inferior;* those with orbits outside Earth's are *superior.*

Any planet can be aligned with both Earth and Sun at two points in its relative orbit. When on the far side of the Sun, the point is called *superior conjunction;* when on the near side, between Earth and Sun (inferior planets only), *inferior conjunction.* The angle between Sun and planet as we see them in the sky is the planet's *elongation.* At both conjunctions the elongation is 0°. *Maximum elongation* is the greatest possible angular distance of the planet from the Sun. An inferior planet must have an elongation of less than 90°. Superior planets can be at any elongation up to 180° (directly opposite the Sun). When at 180° elongation, a planet is at *opposition.* When at 90° elongation, a superior planet is at *quadrature.*

In the illustrations, the successive positions of Earth and representative inferior and superior planets are shown at equal time intervals, and the line of sight connecting them is projected on the celestial sphere. Because both Earth and the other planet are in continuous motion at different speeds, the track of the other planets through the sky may be complicated. Although the orbital motion of all planets is west to east, planets can catch up and pass one another in the sky, seeming to move east to west in what is called *retrograde motion.* Apparent west-to-east motion is called *direct.*

An inferior planet (top illustration) moves faster than Earth. While on the far side of the Sun, its apparent motion is *eastward.* When on the near side of the Sun, it moves westward against the background of stars for a while. Inferior planets also have phases, similar to those of the Moon. The distance from Earth to an inferior planet changes greatly during one *synodic period,* or time it takes to go through all the relative positions once; hence the apparent size also changes. Inferior planets are brighter in the crescent phase (because they are nearer) than in the full phase.

A superior planet, being always farther from the Sun than Earth is, moves more slowly. Its motion is mostly eastward, but around the time of opposition, when Earth is passing it, its motion appears retrograde. Its path in the sky over months may be S-, Z-, or loop-shaped. Superior planets are at "full" phase only when at opposition, but never vary much in apparent shape.

The position of a planet in its relative orbit corresponds to the time of night it is up (see diagrams). Planets near opposition rise at sunset and are in the sky all night. Planets near conjunction rise and set about the same time as the Sun.

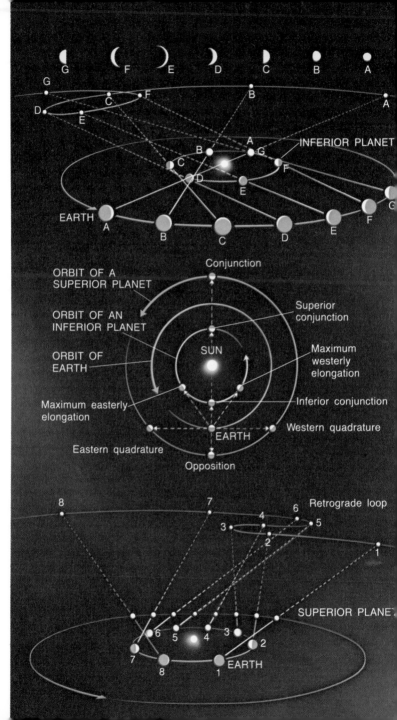

G F E D C B A

INFERIOR PLANET

EARTH

A B C D E F G

Conjunction

ORBIT OF A
SUPERIOR PLANET

ORBIT OF AN
INFERIOR PLANET

ORBIT OF
EARTH

SUN

Superior
conjunction

Maximum
westerly
elongation

Inferior conjunction

Maximum easterly
elongation

Western quadrature

Eastern quadrature

EARTH

Opposition

Retrograde loop

SUPERIOR PLANET

EARTH

THE INVISIBLE UNIVERSE

Most of this book has dealt with the wide variety of sky objects that are visible to the unaided eye, and with explanations of what they are and how they came to be. But on either side of the visible spectrum lies a new universe, most of it not even suspected until this century. It is too vast (or too small), too far or too faint, or too much removed from the visible spectrum, for our unaided eyes to see. Each wavelength of radiation—radio and ultraviolet, x-ray and gamma-ray—gives professional astronomers new information and reveals new kinds of objects to study.

Within our solar system are unseen atomic particles, invisible electrical and magnetic fields, all playing important roles, all affecting what we see and measure on our own planet and others. Several planets—notably giant Jupiter—have extensive invisible yet detectable fields. Spacecraft have revealed dozens of satellites never seen through Earthly telescopes. Some day, quite possibly, we shall find another major planet beyond Pluto.

Our Milky Way galaxy holds newly found objects that only a few decades ago would have been consigned to science fiction. Neutron stars, which are "dying" stars the size of a city, contain a mass equal to the Sun's and have extraordinary gravitational fields. Black holes (which some scientists think don't exist) may explain some of the puzzling observations that have been made of the centers of galaxies. Vast clouds in space, containing such chemicals as water, ammonia, cyanogen, formaldehyde, and alcohol, may have played a role in the formation of life.

Our unaided eyes can see only one major galaxy: M31, in Andromeda. Beyond it, and hence too faint for our eyes but not our telescopes, are billions upon billions of others, as well as clusters of galaxies, and perhaps clusters of clusters. There are vast regions—so far unexplained—where no galaxies are detectable. Each new development in astronomical technology, each attempt to wring more information from the precious bits of light coming across those billions of light-years, allows us to detect even more distant galaxies. We have not yet reached the "edge of the universe," if that term has any meaning at all. We know only approximately how and when the universe came into being with a "Big Bang" about 15 billion years ago. We have less knowledge of how it will end, and when—if end it does.

In the years ahead, new discoveries will undoubtedly make some of our current knowledge obsolete, and some of the "facts" in *Skyguide* will have to be revised. Meanwhile, we must continue learning more about the marvelous universe whose visible appearance this book endeavors to depict and explain. Readers who want to keep up with new developments will find help in the books and magazines listed in the bibliography.

(Right) A radio telescope

POSITIONS OF SUN AND PLANETS, 1990-1997

The table here lists the celestial longitudes of the Sun and naked eye planets. For a definition of celestial longitude, refer to pp. 14-15. To find where, among the constellations, the planets will be, plot the given longitudes on the sky map on the facing page. Remember that the planets can lie as much as a few degrees north or south of the ecliptic, but the positions given are close enough for locating them. The Sun is, of course, always on the ecliptic. Planets are invisible at times because of nearness to the Sun.

	Sun ☉	Mercury ☿	Venus ♀	Mars ♂	Jupiter ♃	Saturn ♄
1990 (May–Dec)						
May 9	48	40	6	344	98	295
May 19	58	38	17	351	100	295
May 29	68	43	28	358	102	295
Jun 8	77	54	40	6	104	295
Jun 18	88	70	52	13	107	294
Jun 28	96	90	64	20	109	293
Jul 8	106	445	76	27	111	293
Jul 18	115	132	88	34	113	292
Jul 28	125	148	100	40	115	291
Aug 7	134	161	112	46	118	290
Aug 17	144	171	124	52	120	290
Aug 27	154	174	136	58	122	289
Sep 6	163	168	148	63	124	289
Sep 16	173	160	161	67	126	289
Sep 26	183	165	173	71	128	289
Oct 6	193	181	186	73	129	289
Oct 16	203	198	198	75	131	289
Oct 26	212	215	211	74	132	289
Nov 5	222	231	223	73	133	290
Nov 15	232	246	236	70	133	291
Nov 25	243	261	248	67	134	292
Dec 5	253	274	261	63	134	293
Dec 15	263	280	274	60	133	294
Dec 25	273	271	286	58	133	295
1991						
Jan 4	283	264	299	58	132	296
Jan 14	293	270	311	59	131	297
Jan 24	304	282	324	61	129	199
Feb 3	314	296	336	64	128	300
Feb 13	324	311	349	67	127	301
Feb 23	334	328	1	71	126	302
Mar 5	344	347	13	76	125	303
Mar 15	354	6	25	80	124	304
Mar 25	4	22	37	85	124	305
Apr 4	14	29	49	90	124	305
Apr 14	24	25	61	96	124	306
Apr 24	33	19	73	102	125	307
May 4	43	19	84	107	125	307
May 14	53	27	96	113	127	307
May 24	62	39	107	119	128	307
Jun 3	72	56	117	125	129	307
Jun 13	82	76	127	130	131	306
Jun 23	91	98	136	136	133	306
Jul 3	101	118	144	142	135	305
Jul 13	110	147	151	149	137	304
Jul 23	120	147	156	155	139	304
Aug 2	129	155	157	161	141	303
Aug 12	139	155	155	167	143	302
Aug 22	149	148	150	174	146	302
Sep 1	158	143	144	180	148	301
Sep 11	168	150	141	186	150	301
Sep 21	178	167	142	193	452	300
Oct 1	187	185	146	200	154	300
Oct 11	197	203	153	206	156	300
Oct 21	207	219	161	213	158	301
Oct 31	217	234	171	220	159	301
Nov 10	227	248	180	227	161	301
Nov 20	237	259	192	234	162	302
Nov 30	247	264	203	241	163	303
Dec 10	258	254	214	248	164	304
Dec 20	268	248	226	255	165	305
Dec 30	278	256	238	262	165	306

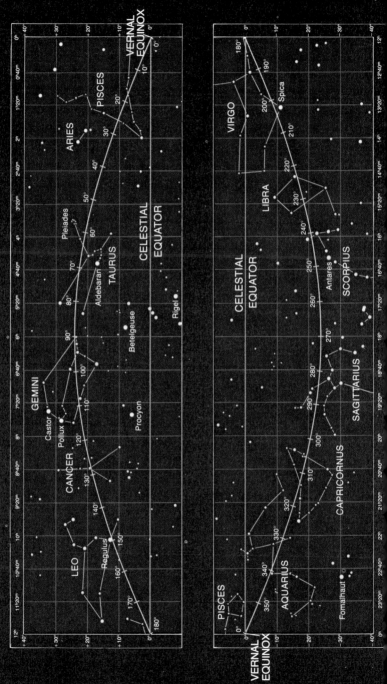

POSITIONS OF SUN AND PLANETS — Contd.

1992

	Sun ☉	Mercury ☿	Venus ♀	Mars ♂	Jupiter ♃	Saturn ♄
Jan 9	288	269	250	270	165	307
Jan 19	298	283	262	277	164	308
Jan 29	308	299	275	285	163	309
Feb 8	319	315	287	292	162	310
Feb 18	329	333	299	300	161	311
Feb 28	339	352	312	308	160	313
Mar 9	349	7	324	315	159	314
Mar 19	359	11	336	323	157	315
Mar 29	9	4	349	331	156	316
Apr 8	19	359	1	339	156	317
Apr 18	28	2	13	346	155	317
Apr 28	38	11	26	354	155	318
May 8	48	25	38	2	155	318
May 18	57	42	50	9	155	319
May 28	67	63	63	17	156	319
Jun 7	77	84	75	24	156	319
Jun 17	86	104	87	32	158	318
Jun 27	96	120	99	39	159	318
Jul 7	105	131	112	46	161	317
Jul 17	115	137	124	53	162	317
Jul 27	124	136	136	60	164	316
Aug 6	134	129	149	67	166	315
Aug 16	143	127	161	73	168	315
Aug 26	153	136	173	80	170	314
Sep 5	163	153	185	86	172	313
Sep 15	173	172	198	92	175	313
Sep 25	182	190	210	97	177	312
Oct 5	192	206	222	102	179	312
Oct 15	202	221	234	107	181	312
Oct 25	212	235	246	111	183	312
Nov 4	222	245	259	114	185	312
Nov 14	232	248	271	116	187	313
Nov 24	242	237	283	118	189	313
Dec 4	252	233	295	118	190	314
Dec 14	262	242	306	116	192	315
Dec 24	273	256	318	113	193	316

1993

	Sun ☉	Mercury ☿	Venus ♀	Mars ♂	Jupiter ♃	Saturn ♄
Jan 3	283	271	329	110	194	317
Jan 13	293	286	340	106	194	318
Jan 23	303	303	350	102	195	319
Feb 2	313	320	0	100	195	320
Feb 12	323	338	8	99	194	321
Feb 22	333	352	15	99	194	323
Mar 4	344	353	19	100	193	324
Mar 14	354	344	20	103	192	325
Mar 24	3	340	17	106	191	326
Apr 3	13	346	11	109	189	327
Apr 13	23	357	6	113	188	328
Apr 23	33	11	4	118	187	329
May 3	43	28	6	123	186	329
May 13	52	49	11	128	185	330
May 23	62	70	18	133	185	330
Jun 2	72	90	26	138	185	330
Jun 12	81	105	35	144	185	330
Jun 22	91	115	45	159	185	330
Jul 2	100	118	56	155	186	330
Jul 12	110	115	67	161	187	330
Jul 22	119	109	78	167	188	329
Aug 1	129	110	89	173	190	328
Aug 11	138	121	101	179	191	328
Aug 21	148	140	112	186	193	327
Aug 31	158	159	124	192	195	326
Sep 10	167	178	136	199	197	326
Sep 20	177	194	148	205	199	325
Sep 30	187	209	160	212	201	324
Oct 10	197	221	173	218	203	324
Oct 20	207	231	185	226	205	324
Oct 30	217	231	198	233	208	324
Nov 9	227	220	210	240	210	324
Nov 19	238	218	223	247	212	324
Nov 29	247	229	235	254	214	325
Dec 9	257	243	248	262	216	325
Dec 19	267	258	260	269	217	325
Dec 29	277	274	272	277	219	327

POSITIONS OF SUN AND PLANETS — Contd.

	Sun ⊙	Mercury ☿	Venus ♀	Mars ♂	Jupiter ♃	Saturn ♄
1994						
Jan 8	288	290	285	285	221	328
Jan 18	298	307	298	292	222	329
Jan 28	308	324	310	300	223	330
Feb 7	318	336	323	308	224	331
Feb 17	328	335	336	316	225	332
Feb 27	338	325	348	323	225	334
Mar 9	348	323	1	331	225	335
Mar 19	358	331	13	339	224	336
Mar 29	8	342	25	347	223	337
Apr 8	18	357	38	355	223	338
Apr 18	28	15	50	3	221	339
Apr 28	38	35	62	10	220	340
May 8	47	56	74	18	219	341
May 18	57	75	86	26	218	341
May 28	67	89	98	33	217	342
Jun 7	76	97	110	41	216	342
Jun 17	86	98	122	48	215	342
Jun 27	95	93	134	55	215	342
Jul 7	105	89	145	62	215	342
Jul 17	114	94	157	69	215	342
Jul 27	124	107	168	76	216	342
Aug 6	133	126	178	83	217	341
Aug 16	143	146	189	90	218	340
Aug 26	153	165	199	96	219	340
Sep 5	162	181	207	102	221	339
Sep 15	172	196	216	109	222	338
Sep 25	182	208	222	115	224	337
Oct 5	192	216	227	120	226	337
Oct 15	201	214	228	126	228	336
Oct 25	211	203	225	131	230	336
Nov 4	221	203	220	136	232	336
Nov 14	231	215	214	140	234	336
Nov 24	242	230	213	144	237	336
Dec 4	252	246	215	148	239	336
Dec 14	262	262	220	150	241	337
Dec 24	272	278	227	152	243	337

	Sun ⊙	Mercury ☿	Venus ♀	Mars ♂	Jupiter ♃	Saturn ♄
1995						
Jan 3	282	294	236	153	245	338
Jan 13	292	309	246	152	247	339
Jan 23	30	320	256	150	249	340
Feb 2	313	317	267	147	251	341
Feb 12	323	307	278	143	252	342
Feb 22	333	307	290	139	253	344
Mar 4	343	316	301	136	254	345
Mar 14	353	329	313	134	255	346
Mar 24	3	344	325	133	255	347
Apr 3	13	1	337	134	255	348
Apr 13	23	21	349	135	255	350
Apr 23	32	43	1	137	255	351
May 3	42	61	13	141	254	352
May 13	52	73	25	145	253	352
May 23	61	78	37	149	252	353
Jun 2	71	76	50	153	251	354
Jun 12	81	71	62	158	249	354
Jun 22	90	71	74	164	248	355
Jul 2	100	78	86	169	247	355
Jul 12	109	92	98	175	246	355
Jul 22	119	112	111	180	246	355
Aug 1	128	133	123	186	246	354
Aug 11	138	152	135	193	246	354
Aug 21	148	168	148	199	246	353
Aug 31	157	183	160	205	247	353
Sep 10	167	194	172	212	248	352
Sep 20	177	200	185	219	249	351
Sep 30	186	197	197	225	250	350
Oct 10	196	187	210	232	252	350
Oct 20	206	188	222	239	254	349
Oct 30	216	201	235	247	256	349
Nov 9	226	218	247	254	257	348
Nov 19	236	234	259	261	260	348
Nov 29	246	250	271	269	262	348
Dec 9	256	265	284	276	264	348
Dec 19	267	281	297	284	267	349
Dec 29	276	295	309	292	269	349

	Sun ☉	Mercury ☿	Venus ♀	Mars ♂	Jupiter ♃	Saturn ♄		Sun ☉	Mercury ☿	Venus ♀	Mars ♂	Jupiter ♃	Saturn ♄
	1996							**1997**					
Jan 8	287	305	321	300	271	350	Jan 2	282	282	260	180	295	1
Jan 18	297	299	333	308	273	351	Jan 12	292	273	272	183	298	2
Jan 28	307	290	346	315	275	352	Jan 22	302	278	285	285	300	3
Feb 7	318	292	358	303	278	353	Feb 1	312	289	297	186	302	4
Feb 17	328	302	9	331	279	354	Feb 11	322	303	310	186	305	5
Feb 27	338	316	21	339	281	355	Feb 21	332	318	322	185	307	6
Mar 8	348	331	32	347	283	356	Mar 3	342	335	335	182	309	7
Mar 18	357	348	43	355	284	358	Mar 13	352	355	347	178	311	8
Mar 28	7	7	53	3	286	359	Mar 23	2	14	0	175	313	9
Apr 7	17	28	63	10	287	0	Apr 2	12	31	12	171	315	11
Apr 17	27	46	72	18	287	1	Apr 12	22	39	25	168	317	12
Apr 27	37	57	80	26	288	2	Apr 22	32	37	37	167	318	13
May 7	47	58	85	33	288	3	May 2	42	31	49	167	320	14
May 17	56	53	88	41	287	4	May 12	51	30	62	168	321	15
May 27	66	50	87	48	287	5	May 22	61	36	74	170	321	16
Jun 6	76	53	83	55	286	6	Jun 1	71	47	86	173	322	17
Jun 16	85	63	77	62	285	7	Jun 11	80	64	98	177	322	18
Jun 26	95	78	73	70	284	7	Jun 21	90	84	111	181	322	19
Jul 6	104	98	72	76	283	7	Jul 1	99	106	123	185	321	20
Jul 16	114	119	75	83	281	7	Jul 11	109	125	135	190	320	20
Jul 26	123	139	81	90	280	7	Jul 21	118	142	147	196	319	20
Aug 5	133	155	88	97	279	7	Jul 31	128	155	159	201	318	20
Aug 15	142	169	97	103	278	7	Aug 10	137	164	171	207	317	20
Aug 25	152	179	106	110	278	6	Aug 20	147	166	183	213	316	20
Sep 4	162	184	117	116	278	6	Aug 30	157	160	195	220	314	20
Sep 14	172	179	127	123	278	5	Sep 9	166	153	206	226	313	19
Sep 24	181	170	138	129	279	4	Sep 19	176	159	218	233	313	19
Oct 4	191	173	150	135	279	3	Sep 29	186	174	229	240	312	18
Oct 14	201	188	162	141	280	3	Oct 9	196	192	241	247	312	17
Oct 24	211	205	173	147	282	2	Oct 19	206	209	261	254	312	16
Nov 3	221	222	185	152	283	1	Oct 29	216	225	262	262	313	15
Nov 13	231	237	198	157	285	1	Nov 8	226	240	273	269	314	15
Nov 23	241	253	210	163	287	1	Nov 18	236	255	282	277	315	14
Dec 3	251	268	222	167	289	1	Nov 28	246	267	291	284	316	14
Dec 13	261	281	235	172	291	1	Dec 8	256	273	298	292	318	14
Dec 23	271	289	247	176	293	1	Dec 18	266	265	302	300	320	14
							Dec 28	276	257	304	308	321	14

PLANETARY SATELLITES DATA

Name	Diameter (km)	Mass (Compared to Moon)	Distance from Planet (1,000 km.)	Period of Revolution (days)	Year First Seen
Earth's Satellite:					
Moon	3,476	1	384.5	27.322	–
Mars' Satellites:					
Phobos	21	0.00000018	9.4	0.319	1877
Deimos	12	0.000000024	23.5	1.263	1877
Jupiter's Satellites:					
Metis	40?	?	128	0.294	1979
Adrastea	25?	?	129	0.297	1979
Amalthea	170	?	112	0.498	1892
Thebe	100	?	222	0.674	1979
Io	3,630	1.214	422	1.769	1610
Europa	3,140	0.663	671	3.551	1610
Ganymede	5,260	2.027	1,070	7.155	1610
Callisto	4,800	1.463	1,885	16.689	1610
Leda	15?	?	11,110	240	1974
Himalia	185	?	11,470	251	1904
Lysithea	35?	?	11,710	260	1938
Elara	75	?	11,740	260	1905
Ananke	30?	?	21,200	631r	1951
Carme	40?	?	22,350	692r	1938
Pasiphae	50?	?	23,330	735r	1908
Sinope	35?	?	23,370	758r	1914
Saturn's Satellites:					
Atlas	30	?	137	0.601	1980
Prometheus	100	?	139	0.613	1980
Pandora	90	?	142	0.628	1980
Janus	190	?	151	0.695	1966
Epimetheus	120	?	151	0.695	1966
Mimas	390	0.0005	187	0.942	1789
Enceladus	500	0.0011	238	1.370	1789
Tethys	1,060	0.0100	295	1.888	1684
Telesto	25	?	295	1.888	1980
Calypso	25	?	295	1.888	1980
Dione	1,120	0.0140	378	2.737	1684
Helene	30	?	378	2.737	1980
Rhea	1,530	0.0338	526	4.517	1672
Titan	5,550	1.831	1,221	15.945	1655
Hyperion	255	?	1,481	21.276	1848
Iapetus	1,460	0.026	3,561	79.331	1671
Phoebe	220	?	12,960	550.46r	1898
Uranus' Satellites:					
Cordelia	25	?	49.8	0.333	1986
Ophelia	30	?	53.8	0.375	1986
Bianca	40	?	59.2	0.433	1986
Cressida	60	?	61.8	0.463	1986
Desdemona	55	?	62.6	0.475	1986
Juliet	85	?	64.4	0.492	1986
Portia	110	?	66.1	0.513	1986
Rosalind	55	?	70.0	0.558	1986
Belinda	65	?	75.3	0.621	1986
Puck	155	?	86.0	0.763	1986
Miranda	485	0.0010	129.9	1.413	1948
Ariel	1,160	0.0182	190.9	2.521	1851
Umbriel	1,190	0.0173	266.0	4.146	1851
Titania	1,610	0.0472	436.3	8.704	1787
Oberon	1,550	0.0397	583.4	13.463	1787
Neptune's Satellites:					
1989 N6*	50	?	48.2	0.296	1989
1989 N5*	90	?	50.0	0.313	1989
1989 N3*	140	?	52.5	0.333	1989
1989 N4*	160	?	62.0	0.396	1989
1989 N2*	200	?	73.6	0.554	1989
1989 N1*	420	?	117.6	1.121	1989
Triton	2,720	0.29	354	5.877	1846
Nereid	300	?	5,600	365.21	1949
Pluto's Satellite:					
Charon	1,300	0.02(?)	19.1	6.387	1978

Note: In the period data, an "r" indicates a retrograde orbit.
*Temporary name until International Astronomical Union decides on permanent name.

PHASES OF THE MOON, 1990-1997

Data in the table are given in Greenwich Mean Time. To convert to time zones in use in the United States, from the times given subtract:

4h to get Eastern Daylight Time

5h to get Eastern Standard or Central Daylight Time

6h to get Central Standard or Mountain Daylight Time

7h to get Mountain Standard or Pacific Daylight Time

8h to get Pacific Standard Time

If the number of hours to be subtracted is greater than the number of hours in the table, add 24h to the time in the table before subtracting. In that case, the date of the event is one day before the date given in the table. For example, if a full moon occurs at 04h30m GMT on June 15, it will occur at 23h30m EST on June 14. Simply remember what time zone you are in when making your calculation. For conversion to other zones, see pp. 18-19.

MOON PHASES

New Moon	First Quarter	Full Moon	Last Quarter
1990			
d h m	d h m	d h m	d h m
Jun 22 18 55	Jun 29 22 07	Jul 8 01 23	Jul 15 11 04
Jul 22 02 54	Jul 29 14 01	Aug 6 14 19	Aug 13 15 54
Aug 20 12 39	Aug 28 07 34	Sep 5 01 46	Sep 11 20 53
Sep 19 00 46	Sep 27 02 06	Oct 4 12 02	Oct 11 03 31
Oct 18 15 37	Oct 26 20 26	Nov 2 21 48	Nov 9 13 02
Nov 17 09 05	Nov 25 13 11	Dec 2 07 50	Dec 9 02 04
Dec 17 04 22	Dec 25 03 16	Dec 31 18 35	
1991			
			Jan 7 18 35
Jan 15 23 50	Jan 23 14 21	Jan 30 06 10	Feb 6 13 52
Feb 14 17 32	Feb 21 22 58	Feb 28 18 25	Mar 8 10 32
Mar 16 08 10	Mar 23 06 03	Mar 30 07 17	Apr 7 06 45
Apr 14 19 38	Apr 21 12 39	Apr 28 20 58	May 7 00 46
May 14 04 36	May 20 19 46	May 28 11 37	Jun 5 15 30
Jun 12 12 06	Jun 19 04 19	Jun 27 02 58	Jul 5 02 50
Jul 11 19 06	Jul 18 15 11	Jul 26 18 24	Aug 3 11 25
Aug 10 02 28	Aug 17 05 01	Aug 25 09 07	Sep 1 18 16
Sep 8 11 01	Sep 15 22 01	Sep 23 22 40	Oct 1 00 30
Oct 7 21 39	Oct 15 17 33	Oct 23 11 08	Oct 30 07 10
Nov 6 11 11	Nov 14 14 02	Nov 21 22 56	Nov 28 15 21
Dec 6 03 56	Dec 14 09 32	Dec 21 10 23	Dec 28 01 55

258

New Moon	First Quarter	Full Moon	Last Quarter

1992

	d h m		d h m		d h m		d h m
Jan	4 23 10	Jan	13 02 32	Jan	19 21 28	Jan	26 15 27
Feb	3 19 00	Feb	11 16 15	Feb	18 08 04	Feb	25 07 56
Mar	4 13 22	Mar	12 02 36	Mar	18 18 18	Mar	26 02 30
Apr	3 05 01	Apr	10 10 06	Apr	17 04 42	Apr	24 21 40
May	2 17 44	May	9 15 43	May	16 16 03	May	24 15 53
Jun	1 03 57	Jun	7 20 47	Jun	15 04 50	Jun	23 08 11
Jun	30 12 18	Jul	7 02 43	Jul	14 19 06	Jul	22 22 12
Jul	29 19 35	Aug	5 10 58	Aug	13 10 27	Aug	21 10 01
Aug	28 02 42	Sep	3 22 39	Sep	12 02 17	Sep	19 19 53
Sep	26 10 40	Oct	3 14 12	Oct	11 18 03	Oct	19 04 12
Oct	25 20 34	Nov	2 09 11	Nov	10 09 20	Nov	17 11 39
Nov	24 09 11	Dec	2 06 17	Dec	9 23 41	Dec	16 19 13
Dec	24 00 43						

1993

	d h m		d h m		d h m		d h m
		Jan	1 03 38	Jan	8 12 37	Jan	15 04 01
Jan	22 18 27	Jan	30 23 20	Feb	6 23 55	Feb	13 14 57
Feb	21 13 05	Mar	1 15 46	Mar	8 09 46	Mar	15 04 16
Mar	23 07 14	Mar	31 04 10	Apr	6 18 43	Apr	13 19 39
Apr	21 23 49	Apr	29 12 40	May	6 03 34	May	13 12 20
May	21 14 06	May	28 18 21	Jun	4 13 02	Jun	12 05 36
Jun	20 01 52	Jun	26 22 43	Jul	3 23 45	Jul	11 22 49
Jul	19 11 24	Jul	26 03 25	Aug	2 12 10	Aug	10 15 19
Aug	17 19 28	Aug	24 09 57	Sep	1 02 33	Sep	9 06 26
Sep	16 03 10	Sep	22 19 32	Sep	30 18 54	Oct	8 19 35
Oct	15 11 36	Oct	22 08 52	Oct	30 12 38	Nov	7 06 36
Nov	13 21 34	Nov	21 02 03	Nov	29 06 31	Dec	6 15 49
Dec	13 09 27	Dec	20 22 26	Dec	28 23 05		

1994

	d h m		d h m		d h m		d h m
						Jan	5 00 00
Jan	11 23 10	Jan	19 20 27	Jan	27 13 23	Feb	3 08 06
Feb	10 14 30	Feb	18 17 47	Feb	26 01 15	Mar	4 16 53
Mar	12 07 05	Mar	20 12 14	Mar	27 11 09	Apr	3 02 55
Apr	11 00 17	Apr	19 02 34	Apr	25 19 45	May	2 14 32
May	10 17 07	May	18 12 50	May	25 03 39	Jun	1 04 02
Jun	9 08 26	Jun	16 19 56	Jun	23 11 33	Jun	30 19 31
Jul	8 21 37	Jul	16 01 12	Jul	22 20 16	Jul	30 12 40
Aug	7 08 45	Aug	14 05 57	Aug	21 06 47	Aug	29 06 41
Sep	5 18 33	Sep	12 11 34	Sep	19 20 00	Sep	28 00 23
Oct	5 03 55	Oct	11 19 17	Oct	19 12 18	Oct	27 16 44
Nov	3 13 35	Nov	10 06 14	Nov	18 06 57	Nov	26 07 04
Dec	2 23 54	Dec	9 21 06	Dec	18 02 17	Dec	25 19 06

MOON PHASES—Contd.

New Moon	First Quarter	Full Moon	Last Quarter

1995

New Moon			First Quarter			Full Moon			Last Quarter		
	d	h	m	d	h	m	d	h	m	d	h	m

New Moon	First Quarter	Full Moon	Last Quarter
Jan 1 10 56	Jan 8 15 46	Jan 16 20 26	Jan 24 04 58
Jan 30 22 48	Feb 7 12 54	Feb 15 12 15	Feb 22 13 04
Mar 1 11 48	Mar 9 10 14	Mar 17 01 26	Mar 23 20 10
Mar 31 02 09	Apr 8 05 35	Apr 15 12 08	Apr 22 03 18
Apr 29 17 36	May 7 21 44	May 14 20 48	May 21 11 36
May 29 09 27	Jun 6 10 26	Jun 13 04 03	Jun 19 22 01
Jun 28 00 50	Jul 5 20 02	Jul 12 10 49	Jul 19 11 10
Jul 27 15 13	Aug 4 03 16	Aug 10 18 15	Aug 18 03 04
Aug 26 04 31	Sep 2 09 03	Sep 9 03 57	Sep 16 21 09
Sep 24 16 55	Oct 1 14 36	Oct 8 15 52	Oct 16 16 26
Oct 24 04 36	Oct 30 21 17	Nov 7 07 20	Nov 15 11 40
Nov 22 15 43	Nov 29 06 28	Dec 7 01 27	Dec 15 05 31
Dec 22 02 22	Dec 28 19 06		

1996

New Moon	First Quarter	Full Moon	Last Quarter
		Jan 5 20 51	Jan 13 20 45
Jan 20 12 50	Jan 27 11 14	Feb 4 15 58	Feb 12 08 37
Feb 18 23 30	Feb 26 05 52	Mar 5 09 23	Mar 12 17 15
Mar 19 10 45	Mar 27 01 31	Apr 4 00 07	Apr 10 23 36
Apr 17 22 49	Apr 25 20 40	May 3 11 48	May 10 05 04
May 17 11 46	May 25 14 13	Jun 1 20 47	Jun 8 11 05
Jun 16 01 36	Jun 24 05 23	Jul 1 03 58	Jul 7 18 55
Jul 15 16 15	Jul 23 17 49	Jul 30 10 35	Aug 6 05 25
Aug 14 07 34	Aug 22 03 36	Aug 28 17 52	Sep 4 19 06
Sep 12 23 07	Sep 20 11 23	Sep 27 02 51	Oct 4 12 04
Oct 12 14 14	Oct 19 18 09	Oct 26 14 11	Nov 3 07 50
Nov 11 04 16	Nov 18 01 09	Nov 25 04 10	Dec 3 05 06
Dec 10 16 56	Dec 17 09 31	Dec 24 20 41	

1997

New Moon	First Quarter	Full Moon	Last Quarter
			Jan 2 01 45
Jan 9 04 26	Jan 15 20 02	Jan 23 15 11	Jan 31 19 40
Feb 7 15 06	Feb 14 08 57	Feb 22 10 27	Mar 2 09 37
Mar 9 01 15	Mar 16 00 06	Mar 24 04 45	Mar 31 19 38
Apr 7 11 02	Apr 14 17 00	Apr 22 20 33	Apr 30 02 37
May 6 20 46	May 14 10 55	May 22 09 13	May 29 07 51
Jun 5 07 03	Jun 13 04 51	Jun 20 19 09	Jun 27 12 42
Jul 4 18 40	Jul 12 21 44	Jul 20 03 20	Jul 26 18 28
Aug 3 08 14	Aug 11 12 42	Aug 18 10 55	Aug 25 02 23
Sep 1 23 52	Sep 10 01 31	Sep 16 18 50	Sep 23 13 35
Oct 1 16 51	Oct 9 12 22	Oct 16 03 46	Oct 23 04 48
Oct 31 10 01	Nov 7 21 43	Nov 14 14 12	Nov 21 23 58
Nov 30 02 14	Dec 7 06 09	Dec 14 02 37	Dec 21 21 43
Dec 29 16 56			

TIMES OF SUNSET AND SUNRISE

Data in the following tables is given in approximate Local Mean Time (LMT). To convert to your zone time, follow the instructions in the next-to-last paragraph on page 18.

SUNRISE

Local mean time of sunrise — Meridian of Greenwich

Date	Lat.	+10°	+20°	+30°	+35°	+40°	+45°	+50°	+60°
		h m	h m	h m	h m	h m	h m	h m	h m
Jan	5	6 18	6 36	6 57	7 09	7 22	7 38	7 58	9 00
	10	6 20	6 37	6 57	7 09	7 22	7 37	7 56	8 56
	15	6 21	6 38	6 57	7 08	7 20	7 35	7 53	8 49
	20	6 22	6 38	6 56	7 06	7 18	7 32	7 49	8 41
	25	6 23	6 37	6 54	7 04	7 15	7 28	7 44	8 31
	30	6 23	6 36	6 52	7 01	7 11	7 23	7 38	8 20
Feb	4	6 22	6 35	6 49	6 57	7 07	7 17	7 30	8 09
	9	6 21	6 33	6 46	6 53	7 01	7 11	7 23	7 56
	14	6 20	6 30	6 42	6 48	6 55	7 04	7 14	7 43
	19	6 19	6 27	6 37	6 43	6 49	6 56	7 05	7 30
	24	6 17	6 24	6 32	6 37	6 42	6 48	6 55	7 16
	29	6 15	6 21	6 27	6 31	6 35	6 40	6 45	7 01
Mar	5	6 12	6 17	6 22	6 24	6 27	6 31	6 35	6 46
	10	6 10	6 13	6 16	6 18	6 19	6 22	6 24	6 32
	15	6 07	6 08	6 10	6 11	6 12	6 12	6 14	6 17
	20	6 04	6 04	6 04	6 04	6 03	6 03	6 03	6 01
	25	6 01	6 00	5 58	5 57	5 55	5 54	5 52	5 46
	30	5 58	5 55	5 52	5 50	5 47	5 44	5 41	5 31
Apr	4	5 56	5 51	5 46	5 43	5 39	5 35	5 30	5 16
	9	5 53	5 47	5 40	5 36	5 31	5 26	5 20	5 01
	14	5 50	5 43	5 34	5 29	5 24	5 17	5 09	4 46
	19	5 48	5 39	5 29	5 23	5 16	5 08	4 59	4 32
	24	5 45	5 35	5 24	5 17	5 09	5 00	4 49	4 17
	29	5 43	5 32	5 19	5 11	5 03	4 52	4 40	4 03
May	4	5 42	5 29	5 15	5 06	4 56	4 45	4 31	3 50
	9	5 40	5 26	5 11	5 01	4 51	4 38	4 23	3 37
	14	5 39	5 24	5 07	4 57	4 46	4 32	4 15	3 24
	19	5 38	5 22	5 04	4 54	4 41	4 27	4 09	3 13
	24	5 38	5 21	5 02	4 51	4 38	4 22	4 03	3 03
	29	5 38	5 20	5 00	4 48	4 35	4 18	3 58	2 54
June	3	5 38	5 20	4 59	4 47	4 32	4 15	3 54	2 46
	8	5 38	5 20	4 58	4 46	4 31	4 14	3 52	2 41
	13	5 39	5 20	4 58	4 45	4 30	4 13	3 50	2 37
	18	5 40	5 21	4 59	4 46	4 31	4 13	3 50	2 35
	23	5 41	5 22	5 00	4 47	4 32	4 14	3 51	2 36
	28	5 42	5 23	5 01	4 48	4 33	4 16	3 53	2 39
July	3	5 43	5 25	5 03	4 50	4 36	4 18	3 56	2 44
	8	5 45	5 26	5 05	4 53	4 39	4 22	4 00	2 51

SUNSET

Local mean time of sunset — Meridian of Greenwich

Date	Lat.	+10°	+20°	+30°	+35°	+40°	+45°	+50°	+60°
		h m	h m	h m	h m	h m	h m	h m	h m
Jan	5	17 52	17 34	17 14	17 02	16 48	16 32	16 12	15 10
	10	17 55	17 37	17 18	17 06	16 53	16 38	16 18	15 19
	15	17 57	17 41	17 22	17 11	16 58	16 44	16 25	15 30
	20	18 00	17 44	17 26	17 16	17 04	16 50	16 33	15 42
	25	18 02	17 47	17 30	17 21	17 10	16 57	16 41	15 54
	30	18 04	17 50	17 35	17 26	17 16	17 04	16 49	16 07
Feb	4	18 06	17 53	17 39	17 31	17 22	17 11	16 58	16 20
	9	18 07	17 56	17 43	17 36	17 28	17 18	17 07	16 33
	14	18 08	17 58	17 47	17 41	17 34	17 25	17 15	16 46
	19	18 09	18 01	17 51	17 46	17 39	17 32	17 24	16 59
	24	18 10	18 03	17 55	17 50	17 45	17 39	17 32	17 12
	29	18 10	18 05	17 58	17 55	17 51	17 46	17 41	17 25
Mar	5	18 11	18 07	18 02	17 59	17 56	17 53	17 49	17 38
	10	18 11	18 08	18 05	18 04	18 02	18 00	17 57	17 50
	15	18 11	18 10	18 08	18 08	18 07	18 06	18 05	18 03
	20	18 11	18 11	18 11	18 12	18 12	18 13	18 13	18 15
	25	18 11	18 12	18 15	18 16	18 17	18 19	18 21	18 27
	30	18 11	18 14	18 18	18 20	18 22	18 25	18 29	18 39
Apr	4	18 10	18 15	18 21	18 24	18 27	18 32	18 37	18 51
	9	18 10	18 17	18 24	18 28	18 32	18 38	18 45	19 04
	14	18 10	18 18	18 27	18 32	18 37	18 44	18 52	19 16
	19	18 11	18 20	18 30	18 36	18 43	18 51	19 00	19 28
	24	18 11	18 21	18 33	18 40	18 48	18 57	19 08	19 41
	29	18 11	18 23	18 36	18 44	18 53	19 03	19 16	19 53
May	4	18 12	18 25	18 39	18 48	18 58	19 09	19 23	20 06
	9	18 13	18 27	18 43	18 52	19 03	19 15	19 31	20 18
	14	18 14	18 29	18 46	18 56	19 07	19 21	19 38	20 30
	19	18 15	18 31	18 49	19 00	19 12	19 27	19 45	20 41
	24	18 16	18 33	18 52	19 03	19 17	19 32	19 51	20 52
	29	18 17	18 35	18 55	19 07	19 21	19 37	19 57	21 02
June	3	18 18	18 37	18 57	19 10	19 24	19 41	20 02	21 11
	8	18 20	18 38	19 00	19 13	19 27	19 45	20 07	21 18
	13	18 21	18 40	19 02	19 15	19 30	19 48	20 10	21 24
	18	18 22	18 41	19 03	19 16	19 32	19 50	20 12	21 27
	23	18 23	18 42	19 04	19 18	19 33	19 51	20 13	21 28
	28	18 24	18 43	19 05	19 18	19 33	19 51	20 13	21 27
July	3	18 25	18 44	19 05	19 18	19 32	19 50	20 12	21 23
	8	18 25	18 43	19 04	19 17	19 31	19 48	20 09	21 18

262

Local mean time of sunrise — Meridian of Greenwich

Date	Lat.	+10°	+20°	+30°	+35°	+40°	+45°	+50°	+60°
		h m	h m	h m	h m	h m	h m	h m	h m
July	13	5 46	5 28	5 08	4 56	4 42	4 26	4 05	3 00
	18	5 47	5 30	5 11	4 59	4 46	4 30	4 11	3 10
	23	5 48	5 32	5 14	5 03	4 50	4 35	4 17	3 20
	28	5 49	5 34	5 17	5 06	4 55	4 41	4 24	3 31
Aug	2	5 50	5 36	5 20	5 10	4 59	4 46	4 31	3 43
	7	5 50	5 37	5 23	5 14	5 04	4 52	4 38	3 55
	12	5 51	5 39	5 26	5 18	5 09	4 58	4 45	4 07
	17	5 51	5 41	5 28	5 21	5 13	5 04	4 53	4 19
	22	5 51	5 42	5 31	5 25	5 18	5 10	5 00	4 31
	27	5 51	5 43	5 34	5 29	5 23	5 16	5 07	4 43
Sep	1	5 51	5 44	5 37	5 33	5 28	5 22	5 15	4 55
	6	5 50	5 45	5 40	5 36	5 32	5 28	5 22	5 07
	11	5 50	5 47	5 42	5 40	5 37	5 34	5 30	5 19
	16	5 50	5 48	5 45	5 44	5 42	5 40	5 37	5 30
	21	5 49	5 49	5 48	5 47	5 47	5 46	5 45	5 42
	26	5 49	5 50	5 51	5 51	5 51	5 52	5 52	5 54
Oct	1	5 49	5 51	5 53	5 55	5 56	5 58	6 00	6 06
	6	5 48	5 52	5 56	5 59	6 01	6 04	6 08	6 18
	11	5 48	5 54	5 59	6 03	6 06	6 11	6 16	6 30
	16	5 49	5 55	6 03	6 07	6 12	6 17	6 24	6 42
	21	5 49	5 57	6 06	6 11	6 17	6 24	6 32	6 54
	26	5 49	5 59	6 09	6 16	6 22	6 30	6 40	7 07
	31	5 50	6 01	6 13	6 20	6 28	6 37	6 48	7 20
Nov	5	5 51	6 03	6 17	6 25	6 34	6 44	6 56	7 33
	10	5 53	6 06	6 21	6 30	6 39	6 51	7 05	7 46
	15	5 55	6 09	6 25	6 35	6 45	6 58	7 13	7 58
	20	5 57	6 12	6 29	6 39	6 51	7 04	7 21	8 11
	25	5 59	6 15	6 33	6 44	6 56	7 11	7 28	8 22
	30	6 01	6 18	6 38	6 49	7 02	7 17	7 36	8 33
Dec	5	6 04	6 21	6 41	6 53	7 07	7 22	7 42	8 43
	10	6 06	6 24	6 45	6 57	7 11	7 27	7 48	8 51
	15	6 09	6 27	6 48	7 01	7 15	7 31	7 52	8 57
	20	6 11	6 30	6 51	7 04	7 18	7 35	7 56	9 02
	25	6 14	6 32	6 54	7 06	7 20	7 37	7 58	9 04
	30	6 16	6 34	6 55	7 08	7 22	7 38	7 59	9 03

263

Local mean time of sunset — Meridian of Greenwich

Date	Lat.	+10°	+20°	+30°	+35°	+40°	+45°	+50°	+60°
		h m	h m	h m	h m	h m	h m	h m	h m
July	13	18 25	18 43	19 03	19 15	19 29	19 45	20 05	21 10
	18	18 25	18 42	19 01	19 13	19 26	19 41	20 01	21 01
	23	18 25	18 41	18 59	19 10	19 22	19 37	19 55	20 51
	28	18 24	18 39	18 56	19 06	19 18	19 31	19 48	20 40
Aug	2	18 23	18 36	18 52	19 02	19 13	19 25	19 41	20 28
	7	18 21	18 34	18 48	18 57	19 07	19 18	19 33	20 15
	12	18 19	18 31	18 44	18 52	19 01	19 11	19 24	20 01
	17	18 17	18 27	18 39	18 46	18 54	19 03	19 14	19 47
	22	18 14	18 23	18 34	18 40	18 47	18 55	19 05	19 33
	27	18 12	18 19	18 28	18 33	18 39	18 46	18 54	19 18
Sep	1	18 09	18 15	18 22	18 27	18 31	18 37	18 44	19 03
	6	18 06	18 11	18 16	18 20	18 23	18 28	18 33	18 48
	11	18 03	18 06	18 10	18 13	18 15	18 18	18 22	18 33
	16	18 00	18 02	18 04	18 05	18 07	18 09	18 11	18 18
	21	17 57	17 57	17 58	17 58	17 59	17 59	18 00	18 03
	26	17 53	17 53	17 52	17 51	17 50	17 50	17 49	17 47
Oct	1	17 50	17 48	17 45	17 44	17 42	17 40	17 38	17 32
	6	17 48	17 44	17 39	17 37	17 34	17 31	17 28	17 17
	11	17 45	17 40	17 34	17 30	17 26	17 22	17 17	17 03
	16	17 42	17 36	17 28	17 24	17 19	17 13	17 07	16 48
	21	17 40	17 32	17 23	17 18	17 12	17 05	16 57	16 34
	26	17 38	17 29	17 18	17 12	17 05	16 57	16 47	16 20
	31	17 37	17 26	17 14	17 07	16 59	16 50	16 38	16 06
Nov	5	17 36	17 24	17 10	17 02	16 53	16 43	16 30	15 54
	10	17 35	17 22	17 07	16 58	16 48	16 37	16 23	15 42
	15	17 35	17 20	17 04	16 54	16 44	16 31	16 16	15 30
	20	17 35	17 19	17 02	16 52	16 40	16 27	16 10	15 20
	25	17 35	17 19	17 01	16 50	16 37	16 23	16 05	15 11
	30	17 36	17 19	17 00	16 49	16 36	16 20	16 02	15 04
Dec	5	17 38	17 20	17 00	16 48	16 35	16 19	15 59	14 58
	10	17 40	17 21	17 01	16 49	16 35	16 18	15 58	14 55
	15	17 42	17 23	17 02	16 50	16 36	16 19	15 58	14 53
	20	17 44	17 25	17 04	16 52	16 38	16 21	16 00	14 54
	25	17 47	17 28	17 07	16 55	16 40	16 24	16 03	14 57
	30	17 49	17 31	17 10	16 58	16 44	16 27	16 07	15 02

SOME MAJOR PLANETARIUMS

For a more extensive list, and more information about planetariums, consult the references given on the following page.

Talbert and Leota Abrams Planetarium
Science Road
Michigan State University
East Lansing, Michigan 48824
(517) 355-4673

Adler Planetarium
1300 South Lake Shore Drive
Chicago, Illinois 60605
(312) 322-0304

American Museum-Hayden Planetarium
81st Street at Central Park West
New York, New York 10024
(212) 873-8828

Buhl Planetarium
Allegheny Square
Pittsburgh, Pennsylvania 15212
(412) 321-4300

Davis Planetarium
Maryland Science Center
601 Light Street
Baltimore, Maryland 21230
(301) 685-2370

R. C. Davis Planetarium
P.O. Box 288
Jackson, Mississippi 39205
(601) 969-6888

Albert Einstein Spacearium
National Air and Space Museum
6th and Independence Avenue
Washington, D.C. 20560
(202) 381-4193

Fels Planetarium
20th and Parkway
Philadelphia, Pennsylvania 19103
(215) 448-1292

Fernbank Science Center
    Planetarium
156 Heaton Park Drive
Atlanta, Georgia 30307
(404) 378-4311

Grace H. Flandrau Planetarium
University of Arizona
Tucson, Arizona 85721
(602) 626-4515

Reuben H. Fleet Space Theater
1875 El Prado
San Diego, California 92103
(714) 238-1233

Gates Planetarium
Colorado Boulevard and
    Montview
Denver, Colorado 80205
(303) 388-4201

Griffith Observatory
2800 East Observatory Road
Los Angeles, California 90027
(213) 664-1181

George T. Hansen Planetarium
15 South State Street
Salt Lake City, Utah 84111
(801) 364-3611

Charles Hayden Planetarium
Museum of Science
Boston, Massachusetts 02114
(617) 723-2500

Louisiana Arts and Science
    Center Planetarium
502 North Boulevard
Baton Rouge, Louisiana 70802
(504) 344-9465

McDonnell Planetarium
5100 Clayton Avenue
St. Louis, Missouri 63110
(314) 535-5811

Morehead Planetarium
University of North Carolina
Chapel Hill, North Carolina
    27514
(919) 933-1237

Morrison Planetarium
California Academy of Sciences
Golden Gate Park
San Francisco, California 94118
(415) 221-5100

Space Transit Planetarium
3280 South Miami Avenue
Miami, Florida 33129
(305) 854-4242

Strasenburgh Planetarium
663 East Avenue
Rochester, New York 14607
(716) 271-4320

Vanderbilt Planetarium
178 Little Neck Road
Centerport, New York 11721
(516) 262-7800

OBSERVATORIES

The following is a list of some observatories in the United States that are occasionally open to the public for lectures, public viewing, or tours. For a more complete list, see *U.S. Observatories: A Directory and Travel Guide* by H. T. Kirby-Smith, Van Nostrand Reinhold Company, New York, 1976.

Allegheny Observatory
Department of Physics and Astronomy
University of Pittsburgh
Pittsburgh, Pennsylvania 15214
(412) 624-4290

Hale Observatories (Mt. Wilson and
   Mt. Palomar)
California Institute of Technology
Pasadena, California 91101

Harvard-Smithsonian Center for
   Astrophysics
"Open Nights"
Center for Astrophysics
60 Garden Street
Cambridge, Massachusetts 02138
(617) 495-7000

Kitt Peak National Observatory
P.O. Box 26732
Tucson, Arizona 85726
(602) 325-9204

Lick Observatory
Department of Astronomy
University of California at Santa Cruz
Santa Cruz, California 95064
(408) 429-2513

McDonald Observatory
Department of Astronomy
University of Texas
Austin, Texas 78712
(512) 471-4462

National Radio and Ionospheric
   Observatory (Puerto Rico)
P. O. Box 995
Arecibo, Puerto Rico 00612
(809) 878-2612

National Radio Astronomy
   Observatory
Charlottesville, Virginia 22901
(804) 296-0211

Sproul Observatory
Swarthmore College
Swarthmore, Pennsylvania 19081
(215) 544-7900 Ext. 207

United States Naval Observatory
Washington, D.C. 20390
(202) 254-4533

Yerkes Observatory
University of Chicago
Chicago, Illinois 60637
(312) 753-8180

BIBLIOGRAPHY
Books

Textbooks

Abell, George O., *Drama of the Universe*, Holt, Rinehart and Winston, New York, NY, 1978

——, *Exploration of the Universe*, 3rd ed., Holt, Rinehart and Winston, New York, NY, 1975

Kaufmann III, William J., *Exploraton of the Solar System*, Macmillan, Inc., New York, NY, 1978

Popular Books on Astronomy

Allen, Richard Hinckley, *Star Names, Their Love and Meaning*, Dover Publications, New York, NY, 1963

Bok, Bart J. and Priscilla F., *The Milky Way*, 4th ed., Harvard University Press, Cambridge, MA, 1974

Brown, Peter Lancaster, *Comets, Meteorites, and Men*, Taplinger Publishing Co., New York, NY, 1974

Clayton, Donald D., *The Dark Night Sky*, Quadrangle Books, New York, NY, 1975

Ferris, Timothy, *The Red Limit*, William Morrow & Co., New York, NY, 1977

Gallant, Roy A., *The Constellations*, Four Winds Press, New York, NY, 1979

——, *National Geographic Picture Atlas of Our Universe*, National Geographic Society, Washington, DC, 1980

Hawking, Stephen W., *A Brief History of Time*, Bantam Books, New York, NY, 1988

Kaufmann III, William J., *Relativity and Cosmology*, 2nd ed., Harper & Row, New York, NY, 1977

Kirby-Smith, H.T., *U.S. Observatories: A Directory and Travel Guide*, Van Nostrand Reinhold, New York, NY, 1976

Krupp, E. C., ed., *In Search of Ancient Astronomies*, McGraw-Hill Book Co., New York, NY, 1979

Ley, Willy, *Watchers of the Skies*, Viking Press, New York, NY, 1963

Lum, Peter, *The Stars in Our Heaven*, Pantheon Books, New York, NY, 1948

Richardson, Robert S., *The Star Lovers*, Macmillan, Inc., New York, NY, 1967

Shipman, Harry L., *Black Holes, Quasars, and the Universe*, 2nd ed., Houghton Mifflin Co., Boston, MA, 1980

Sullivan, Walter, *Black Holes*, Anchor Press, New York, NY, 1979

Waugh, Albert E., *Sundials*, Dover Publications, New York, NY, 1973

Whitney, Charles A., *The Discovery of Our Galaxy*, Alfred A. Knopf, New York, NY, 1971

Books on Observing the Sky

Burnham, Robert, Jr., *Burnham's Celestial Handbook* (in 3 volumes), Dover Publications, New York, NY, 1978

Cherrington, Jr., Ernest H., *Exploring the Moon Through Binoculars*, McGraw-Hill Book Co., New York, NY, 1969

Cleminshaw, Clarence H., *The Beginner's Guide to the Sky*, Thomas Y. Crowell Co., New York, NY, 1977

Government Printing Office, *Smithsonian Astrophysical Observatory Star Catalog* (4 volumes), Washington, DC, 1966

Mayall, R. Newton and Margaret W., *Olcott's Field Book of the Skies*, 4th ed., G. P. Putnam's Sons, New York, NY, 1954

————, and Jerome Wyckoff, *Sky Observer's Guide*, Golden Press, New York, NY, 1977

MIT Press, *Smithsonian Astrophysical Observatory Star Atlas*, Cambridge, MA, 1969

Muirden, James, *The Amateur Astronomer's Handbook*, rev. ed., Thomas Y. Crowell Co., New York, NY, 1974

Norton, Arthur P., *Norton's Star Atlas*, Sky Publishing Corp., Cambridge, MA, 1978

Royal Astronomical Society of Canada, *Observer's Handbook* (annual), Toronto, Canada

Sidgwick, J. B., *Amateur Astronomer's Handbook*, Dover Publications, New York, NY, 1980

U.S. Naval Observatory, *The Astronomical Almanac* (annual), Government Printing Office, Washington, DC

Technical Books

Duffett-Smith, Peter, *Astronomy With Your Personal Computer*, Cambridge University Press, Cambridge, UK, 1985

Eastman Kodak Company, *Astrophotography Basics*, Publication AC-48 (available at most photography stores), Rochester, NY, 1980

McCuskey, Sidney W., *Introduction to Celestial Mechanics*, Addison-Wesley Publishing Co., Reading, MA, 1963

Meeus, Jean, Carl C. Grosjean, and Willy Vanderleen, *Canon of Solar Eclipses*, Pergamon Press, Inc., Elmsford, NY, 1966

Smart, W. M., *Textbook on Spherical Astronomy*, 6th ed., Cambridge University Press, New York, NY, 1977

Magazines

Popular Magazines on Astronomy

Astronomy, AstroMedia Corp., P.O. Box 92788, Milwaukee, WI 53202

Griffith Observer, Griffith Observatory and Planetarium, 2800 East Observatory Rd., Los Angeles, CA 90027

Mercury, Astronomical Society of the Pacific, 390 Ashton Ave., San Francisco, CA 94112

Sky and Telescope, Sky Publishing Corp., 49 Bay State Rd., Cambridge, MA 02138

Technical Magazines with Frequent Articles on Astronomy

Science, American Association for the Advancement of Science, 1515 Massachusetts Ave., N.W., Washington, DC 20005

Scientific American, 415 Madison Ave., New York, NY 10017

SOURCES OF ASTRONOMICAL EQUIPMENT
AND REFERENCES

SOURCE	MATERIALS
Celestron International 2835 Columbia Street Torrance, California 90503	Catadioptric telescopes and related equipment
Criterion Scientific Instruments Inc. 620 Oakwood Avenue West Hartford, Connecticut 06110	Reflecting and catadioptric telescopes and related equipment
Edmund Scientific Corporation Edscorp Building 101 East Gloucester Pike Barrington, New Jersey 08007	A variety of telescopes, equipment, optics, and other science items
Farquhar Globes 5007 Warrington Avenue Philadelphia, Pennsylvania 19143	Celestial globes
Hansen Planetarium Publications 1098 South 200 West Salt Lake City, Utah 84101	Astronomical slides, posters, photographs, other items
Meade Instruments Corporation 1675 Toronto Way Costa Mesa, California 92626	Telescopes and equipment
Questar P.O. Box C New Hope, Pennsylvania 18938	Catadioptric telescopes
S&S Optika, Ltd. 3855 South Broadway Englewood, Colorado 80110	Telescopes and equipment
Sky Publishing Corporation 49 Bay State Road Cambridge, Massachusetts 02138	Star atlases, astronomy books, educational materials, *Sky & Telescope* magazine
Unitron Instruments 175 Express Street Plainview, New York 11803	Refracting telescopes and equipment
Willmann-Bell, Inc. P.O. Box 35025 Richmond, Virginia 23235	Excellent source of books and software
Zephyr Services 1900 Murray Avenue Pittsburgh, Pennsylvania 15217	Astronomy software

INDEX

Most page references below refer to text. Relevant illustrations usually are to be found on the page facing the text.

The first page cited for any technical term listed below usually contains a definition or explanation of that term.

With a few famous exceptions, individual stars are indexed only by proper names. Greek-letter designations can be found in star lists and on constellation charts.

271

274

277

ACKNOWLEDGMENTS

We are indebted to the following individuals and institutions for the photographs used in this book. Mark R. Chartrand: 35b, 251. Lick Observatory: 35t, 59, 61, 63, 211, 213, 215, 217, 233, 235, 237, 239. Dennis Milon: 229, 231. Jerome Wyckoff: 37. Yerkes Observatory: 35m, 57, 227. The table on p. 37 (© Eastman Kodak Company, 1980), has been adapted from Kodak Pamphlet No. AC-48, *Astrophotography Basics*. The quotation on p. 5 is from Volume 12 of *Great Books of the Western World* (Copyright 1982 by Encyclopedia Britannica, Inc.), H. A. J. Monroe's translation of Lucretius' *De Rerum Natura*. Used with permission of Encyclopaedia Britannica.